D1232315

Why Does
My Parrot ...?

Why Does My Parrot ...?

Rosemary Low

Drawings by
Ian Lorriman

Revised and updated

SOUVENIR PRESS

Copyright © 2015 by Rosemary Low

The right of Rosemary Low to be identified as author
of this work has been asserted by her in accordance with
the Copyright, Designs and Patents Act 1988.

First published 2000 by Souvenir Press Ltd,
43 Great Russell Street, London WC1B 3PA
This revised paperback edition 2015

All Rights Reserved. No part of this publication
may be reproduced, stored in a retrieval system,
or transmitted, in any form or by any means, electronic,
mechanical, photocopying, recording or otherwise without
the prior permission of the Copyright owner.

ISBN 9780285643055

Printed and bound by CPI Group (UK) Ltd, Croydon, CR0 4YY

CONTENTS

Introduction 9

PART ONE: BEHAVIOUR THERAPY FOR PARROTS 13

1 What affects behaviour? 15
2 Causes of unacceptable behaviour 36

PART TWO: WHAT IS A PARROT? 47

3 Their lives in the wild 49
4 Parrot species and how they behave 62

PART THREE: WHY DOES MY PARROT ...? 71

A Adolescence 75
Aggression 76
Anthropomorphism 80
Attack 81
B Bach remedies 86
Back, lying on 87
Bathing 87
Beak clicking 89
Beak grinding 89
Biting 89
Blindness and cataracts 92
Blushing 92
Boarding 93
Bonding 94
Buttons 96

C	'Cage-hate'	97
	Children and babies	99
	Coughing	100
	'Cuddly-tame'	100
D	Destructiveness	102
	Dominance	105
E	Euthanasia	107
	Eye contact	107
F	Fear	109
	Feather Plucking	112
	Feeding	117
G	Grieving	120
H	Handling	121
	Hands	122
	'House-training'	123
I	Intelligence	124
J	Jealousy	126
K	Kissing	129
L	Lameness	130
	Laughing	131
	Laying	131
	Loneliness	133
M	Mimicry	135
	Moodiness	137
	Moulting	137
	Music	138
	Mutilation, self	139
N	Nervousness	141
O	Obesity	142
	Old age	143
P	Phobic behaviour	145
	Preening	146
	Predictability	148
	Prevention	149
	Punishment	149
R	Regurgitation	150
	Rings (Bands)	151
S	Screaming	154
	Sexual behaviour	162
	Sexuality – human and psittacine	165

	Shoulder, sitting on	166
	Sleep	168
	Smell, sense of	169
	Smoking	170
	Sneezing	171
	Socialisation	171
	Stereotypic behaviour	173
	Sun	174
T	Tail-bobbing	176
	Talking	176
	Taming	181
	TCM – Traditional Chinese Medicine	184
	Television	185
	Territoriality	187
	Toenails	189
	Toes, mutilation of	191
	Tool-use	192
	Touching	193
	Toys	194
	Training	195
	Tricks	201
V	Veterinary surgeons	204
	Vision, colour	204
W	Wheezing	206
	Whining	206
	Wing-clipping	207
Y	Yawning	215
	Epilogue	217
	References and Sources	219
	Index	221

INTRODUCTION

My friend Irene was travelling across London with her Blue and Yellow Macaw, Max, in a plastic pet carrier. Her journey entailed using the Underground. Max was quiet until the train reached Baker Street. A man got on and enquired in a very loud voice: 'Does this train go to Liverpool Street?' Quick as a flash Max replied in a loud voice: 'Right!' The carriage erupted with human laughter. A girl sitting opposite Irene was reduced to tears, with mascara running down her face.

Max spent the rest of the journey trotting out his favourite phrases, such as 'What's for dinner?' and 'How are you?', interspersing these questions with manic laughter. He even attempted a rendition of 'Getting to know you', one of the few songs he has been able to master. He was so excited at all the attention he was receiving, he came out with practically his entire repertoire. When Irene left the train, Max got a round of applause. As most Londoners know, arousing tube travellers from their state of indifference to the world around them is no easy matter. But parrots fascinate people.

The colours, the watchful eyes, the ability to mimic, the droll mannerisms! Parrots are spell-binding in their beauty and intelligence. The urge to own one can spring from a single encounter, such as that Underground journey. But the reality of owning a parrot is that it is not all fun. It is a huge commitment, and like a commitment to a human, one has to work at it to make it work. Parrots were not designed to live in houses. They are noisy and destructive and suffer probably more than any other animal when kept in an unstimulating environment. They need constant interaction, either with a human or another parrot, to keep them happy and healthy. Keeping a parrot is so much more demanding than keeping a dog or a cat. Alas, the fact that many owners have failed, and failed miserably, is evidenced in the growing number of parrot refuges. They are filled with feather-plucked or phobic parrots whose former owners had no idea of their emotional needs.

Unlike many birds, most parrot species form very strong pair bonds. In a captive situation, where they have no mate of their own species, they will transfer this devotion to a human companion—but only if the human is sympathetic and sensitive to the parrot's personality and needs. Unlike a dog, a parrot does not give its friendship unconditionally. But win the heart of a parrot and you have a friend for life. I know.

My Yellow-fronted Amazon Parrot, Lito, was with me for 39½ years. She crops up often in this book, so let me tell you something about her. She was hatched in the forests of the neotropics, probably near the borders of Colombia and Panama. Almost certainly when she was very young she was removed from the nest (perhaps by felling the tree) and hand-reared by natives. She spent long enough with the family, perhaps even several years, to learn to speak a few words of Spanish. She was well looked after and surely loved, by this family. I know this because her temperament clearly indicates that her early years were happy ones.

One day a man came to the settlement buying birds and animals. (Very few parrots were bred in captivity in those days.) The sum offered for Lito was tempting. It would enable the family to eat well for a few days. Lito was sold. She was taken, probably by boat, with other parrots, to a crude holding centre where the cages were crowded and none too clean, and the food consisted mainly of bananas and rice. After a stay of a few days there she was taken to another place, put in a wooden box and shipped to England.

This was in the 1960s, long before the days when birds arriving in the UK (or anywhere else) had to be quarantined. A dealer would collect the birds from the airport, take them back to his premises and quickly sell them on, perhaps to another dealer. On this day in March 1967, Lito and a number of other parrots, mostly Amazons, had just been collected from the airport. A couple of people watched with interest as the dealer opened the boxes. Out walked Lito with a confident air. In the instant, one of the spectators (me) claimed her for her own.

She was a priceless companion. She has taught me so much about living with a parrot. I hope I can pass on some of this to you, to help you to live happily with a parrot for 39½ years or more.

Part One

BEHAVIOUR THERAPY
FOR PARROTS

1
WHAT AFFECTS BEHAVIOUR?

Many parrot owners are unaware of the various factors which play a part in the behaviour of their pet. Parrots—unlike finches or Canaries, for example—are emotionally quite complex creatures. They also have excellent memories. Both these facts mean that their behaviour can be profoundly influenced by past experiences. Most parrots are flock animals—but in recent years many have been hand-reared for the pet market, denying them the knowledge of their own species from an early age. Emotionally, this can scar them for life, especially if they have been weaned too early. Health and diet are other factors which can influence behaviour. And, as in humans, characteristics of the individual are involved in how they behave. The emotional environment also influences this. Let us now look at these aspects in some detail.

PERSONALITY AND BEHAVIOUR OF THE OWNER

This has an enormous impact on the behaviour of the parrot. It must be said that there are people who should never be allowed near a parrot, let alone own one. Their general demeanour may be loud and threatening. Birds prefer people who are quiet and gentle in movement and manner. Such people tend to be naturally more sympathetic towards living creatures.

Some people have no interest in wildlife or pets, yet they are parrot owners. They might acquire a parrot for the wrong reason. This may be because it is perceived as fashionable, or because walking about with a macaw on the shoulder attracts attention. If a natural empathy with birds does not exist, a parrot can sense that. The larger parrots are extraordinarily sensitive creatures. Unlike many dogs, parrots do not automatically adore their owner. In fact, they may take a strong dislike to someone who has no affinity with birds. The same could occur with someone who is totally devoted to parrots but has the misfortune to remind their parrot of another human being, one who has not dealt with it in a sympathetic manner. It could take a very long time to eradicate these memories—or perhaps they will persist for ever. Some aspects of human behaviour—quick movements and loud voices, for example—can be modified to make the person more acceptable. This also applies to children.

Surprisingly enough, there are people who are afraid of their parrots. Yes, those big beaks can inflict serious injury. But intelligent, sympathetic owners are almost never bitten once a rapport has been established between bird and owner. Parrots can sense or observe fear. They observe it in the hesitant movements of someone offering a hand to step on, for example. While this will usually be ignored by a young hand-reared parrot, it will be acted upon by a parrot with more experience of people.

Just as a parrot can observe when a person is nervous or in a bad mood, the owner should also learn to read his or her parrot's state of mind. Human beings are able to describe their emotions to each other in a sophisticated and complex language. Birds, of course, cannot do this—but if we study a particular species we can understand basic emotions and vocalisations. If we live closely with a pet bird we begin to understand much more about this individual. We should also accept that birds try to communicate with us but unfortunately most people are not receptive.

There are times when a parrot wants to be left alone. The more aggressive species may bite if an attempt is made to handle them then. Grey Parrots are among the least aggressive species. When interacting with their own kind, aggressive behaviour is rare. If, for example, a Grey does not welcome human attention at a particular time, it would be more likely to react by firmly clasping its beak around a finger. Not by biting. But it will bite after it has issued this warning which has been ignored. It will also bite in fear.

They'll never find me in here

How you react to your parrot in response to a certain behaviour from it will, of course, also affect its behaviour. If, for example, when your parrot screams you scream back—in annoyance—the parrot perceives this as a stimulating response and will scream all the more. If when it screams you ignore it totally—not even a glance in its direction—you have done nothing to reinforce its screaming habit. Screaming has not gained your attention, so ultimately, the habit may subside. This assumes that the parrot is screaming to gain your attention; this is the usual cause. However, if it is screaming for some other reason, ignoring it will not be effective.

YOUR MOOD

Be very aware of the fact that your own state of mind can profoundly influence your parrot's behaviour. This is especially true if you become annoyed with him. This will not earn your parrot's respect. Indeed, if he is already playing you up, becoming annoyed will only make matters worse.

One owner recorded: 'Think about the last time you came home upset. Was your Grey Parrot a little more difficult to deal with? Did it refuse to come out of its cage? Did a finger get nipped? Was more food than usual thrown out of the cage? Were there more screams that grated on your nerves? Stress is bad enough for us, but it's far more toxic to our Greys. That's because Greys mirror back our moods. If you're happy, your Grey will be happy; but, if you're angry or nervous, chances are that your Grey will react by being "difficult" (either by biting you or its feathers)' (Zadalis, 1996).

Ellen Zadalis suggested that we must focus on ourselves, identify when we begin to become stressed and start to undo the behaviour. We should do something to relax and to clear our minds. She does so by listening to Louis Armstrong singing 'What a wonderful world'.

As an example of how our birds can react if we become annoyed, the lories in my outdoor aviaries are shut into the house part at night. There are two pairs which sometimes are not too happy about going in. I have to adopt different tactics for each pair. If I became annoyed with the Rajahs for not entering the house, the male became more and more difficult, and occasionally even aggressive. If they are inside when I go to shut them up but move into the outdoor flight when they see or hear me coming, I would have problems if I entered the flight to persuade them into the house. I learned to deal with this problem by ignoring them and entering the house and carrying out some jobs there. They would then come in to see what I was doing and I could run around to the outside flight and shut the hatch. The secret was to avoid the confrontation that occurred if I went into their flight.

A young pair of Stella's Lorikeets reacted quite differently. They had previously behaved well, going inside as soon as I came to shut their hatch. Then they started to challenge me: they did not want to go inside and I had to resort to chasing them around, which was more stressful to me than to them. They are totally tame and fearless. On the third night of this I got the catching net and caught up the male to put him inside. He did not like that. Most parrots hate nets. On subsequent evenings I would take the net with me. I only had to show it to them and they would go straight inside. This ruse would not have been successful with the male Rajah. The sight of the net would have made him aggressive.

I mention these incidents to show that there are ways to avoid confrontations. If you feel yourself getting annoyed, walk away.

HEALTH AND DIET

Dietary deficiencies in parrots are common. They can affect behaviour. A parrot could be sick with, for example, a low-grade bacterial infection which is not immediately life-threatening. Or it could be terminally ill and show no outward sign of this until a few days before it died. In either case, its behaviour could alter because it was feeling unwell or in pain. And pain could also cause it to pluck itself.

No factor has a greater influence on a bird's health and, ultimately, on its lifespan, than diet. But the choice of food is by no means easy. Feed a seed-based diet to certain species which can exist on seed and one risks life-threatening deficiencies, especially of Vitamin A which causes, for example, catarrh (*see* Sneezing).

Grey Parrots are especially susceptible to calcium deficiency when fed a seed diet. (Corticosteroids are sometimes used to treat Greys; but note that they are known to induce hypocalcaemia). This results in fits (seizures) and, if left untreated, early death. So the alternative is to feed pellets which contain sufficient levels of calcium and basic nutrients. Manufacturers claim that these are a complete food. Complete for which species? There are more than 200 parrot species in captivity. In the wild, no two species consume exactly the same range of foods. Obviously, there is no processed food that is equally suitable for so many species with such differing dietary needs.

Those who keep mutation parrots and parrakeets should also be aware of the fact that the metabolism of these birds is different to that of normally coloured parrots. Pelleted foods can have adverse effects on them. It seems that otherwise necessary levels of Vitamin D3, calcium and phosphorus in commercial diets are not well tolerated by certain mutations, such as fallow Cockatiels and dilute parrotlets.

Health has a much greater influence on behaviour than many parrot owners realise. For example, feather plucking is often the result of bacterial infections, pain and allergies.

METHOD OF HAND-REARING

The purchaser of a hand-reared parrot assumes that it will be easy to handle, tame and affectionate—because it is hand-reared. This does not automatically qualify a parrot to make a good pet. The way in which it was hand-reared and weaned can have a profound and long-term effect on its behaviour. Even the method of feeding can be significant. The most caring hand-feeders, and those who feed comparatively small numbers, will use a spoon. This is the most natural method because it is nearest to the way in which a parent feeds its young. It is also the best method from the chick's point of view, as, when it has had enough, it ceases to feed. Syringe-feeding into the mouth comes second in this respect. Both are quite time-consuming, allowing the feeder to establish a relationship with the young parrot which helps it to relate to other humans. Syringe-feeding into the crop is much quicker, so a chick can be handled for a minimum period, as is the case where large numbers are being hand-reared. However, this does not preclude a good relationship between parrot and humans, at this early age, if the feeder is prepared to devote a little extra time to each young parrot. There is a fourth method of giving food to chicks—gavage feeding. A gavage is a metal tube designed to administer medication. A hard object must be very uncomfortable when it is pushed down the throat into the crop. In my opinion, it should never be used for hand-feeding. One purchaser of a gavage-fed young parrot described it as being scared to death of anyone approaching with anything in their hands. It hated meal times. After six months it became more confident, accepting that its new owner was not going to open its beak and force a metal tube down it.

Another factor that can influence the behaviour of hand-reared parrots is whether there is a middle man between breeder and ultimate purchaser. The middle man is usually a pet store owner. Some personnel have no notion of the correct way to feed and interact with parrots at weaning stage. Perhaps, too, there may not be sufficient staff to spend enough time with them. Young parrots might be malnourished or neglected. At that important stage of their lives, this could cause severe behavioural problems. If a hand-reared parrot remained in a store for some weeks after weaning and was not handled on a regular basis, it might become very difficult to hold. Certain species would lose their tameness. The purchaser of a young parrot from a pet store should always ask to handle it. It would be unreasonable to refuse this request, thus an answer in the negative should arouse grave suspicion.

OLD AGE

The behaviour of a young parrot, parrakeet or Cockatiel which has been hand-reared, is very different to that of an adult. There is a strange assumption on the part of many people who see a cuddly young cockatoo or an appealing baby Amazon, that this is how it will be for the rest of its life. In the world of harsh truths it would be appropriate to display a five-year-old of the same species next to the cute baby—and ask: 'Can you cope with the mature counterpart?'

Post-weaning parrots of the species popularly kept as pets are clinging, affectionate, adoring, quiet and compliant. This stage does not last long. Most soon become challenging, noisy and even nippy and unco-operative. This is why it is so important to start training at an early age.

Just as in humans, old age can, of course, also affect behaviour. Old birds are less active; they may become a little bad tempered, especially when handled. Nearly all aged parrots have arthritis in their feet; they should be handled with care. The result is that their grip on the perch is poor and they will fall off the perch at night. Perches should therefore be placed low down in the cage. A vertical ladder-like arrangement of perches might help to prevent falls at night. Arthritis might even affect their wings, making them unable to fly upwards. Very old parrots usually suffer from cataracts. When their sight is impaired, special care will need to be taken when they are free outside the cage.

ENVIRONMENT

The position of a parrot's cage in the home could be crucial to his health and happiness. He needs to feel secure. This can usually be achieved by placing the cage in a corner or in an alcove. The worst position is where the cage can be approached from every side. The height of the cage is also important. The parrot should be at or just below your eye level. If he is above this height, his elevated position will, it appears to him, give him an elevated status. In nature, the dominant birds probably take the highest perches. If he is elevated, he is likely to be much more difficult to control. Likewise, never place a cage where the bird's eye level is below that of your waist. Many perching birds feel insecure when kept permanently at a low level. In nature this would make them most vulnerable to predators. Such a position can result in a lack of self-confidence. If this is the case a parrot would be quiet and unlikely to learn to mimic (*see* also Dominance).

Locations to avoid are those near or facing a television set or computer screen. The flickering lights are very disturbing. Also, the equipment probably emits frequencies which will diminish the bird's sense of well-being. This could affect his behaviour on a permanent basis, making him sluggish or sleepy. It is important that a parrot has enough sleep, especially when young.

Many parrot owners fail to understand that a small cage can have a negative effect on a large parrot's behaviour. Nervous parrots and aggressive cockatoos, or parrots which were wild caught, show a marked decrease in aggression or generally unfriendly behaviour when placed in a small (or large) flight, after being confined to a cage. The large macaws need a lot of headroom and will feel a sense of increased well-being when placed in a cage which has at least 60 cm (two feet) of headroom.

THE INFLUENCE OF COLOUR

An aspect that is seldom considered is that colour can be a negative influence. This applies to the colour of the walls in the area of a parrot's cage and to the colours worn by the people in the vicinity. In nature, blocks of bright colours are very rare; the muted browns of trees, a hundred shades of green, including bright tones which are broken up by light and shade, small points of red and orange which are blossoms, are the tones which parrots see. The brightest colours are those of their own plumage. Walls in stark white and, much worse, solid bright colours, create a harsh, unnatural environment. This would be stressful, at least initially, to wild-caught parrots or those from aviaries.

The very worst colour that a person can wear around birds is bright red. When this is moving (a walking person), a bird can panic with fatal consequences. One day I received a telephone call from a friend who kept a pair of Himalayan Monals, gorgeous pheasants with iridescent plumage. He was very upset as an unfortunate tragedy had just occurred. His partner had gone out to shut the monals in for the night and the male had panicked, hit the aviary wall and broken his neck. Why? This was a nightly ritual. But on that occasion his partner was wearing a bright red jacket.

In nature red is the colour of threat. The males of one species of chameleon which are normally brown, turn brilliant orange-red when they are about to fight another male. I never wear red near birds. Well, almost never. I once had a bright red cardigan which was always covered by a jacket when I went to my aviaries. One day the weather suddenly turned warm and I removed my jacket, forgetting about the red cardigan. When I put my hand into the flight of a pair of Rajah Lories to remove their nectar pot, the male bit me very hard. It was a deep bite—like a razor slash. (Lories have very sharp beaks.) Instantly I realised why. Not only was I wearing red but I was flaunting it in front of a parrot which is mainly black. Its wings are lifted during aggressive encounters to reveal bright red underwings. Solid red is the most threatening sight for this particular species—and instils fear into many others.

Birds have an awareness of colours which is richer than ours. They are also able to see ultra-violet colours. Dr Rosina Sonnenschmidt, who specialises in holistic treatment for parrots, describes what she calls three life colours for birds—green, blue and gold. Sit under a tree on a sunny day and these are the colours you experience when you look upwards, gold being light in constant movement. She believes that blue is extremely important for the welfare of intelligent parrots. She suggests that when a bird is being taught to talk, the teaching sessions should be limited to 15 minutes per day and the bird should be in blue light during this period. Birds which had lost their voice due to shock or immense stress started making sounds as soon as they were put under blue light (Sonnenschmidt, 1996a).

EMOTIONAL ENVIRONMENT

In *Why does my dog. . .?* John Fisher commented: 'There is little doubt that our pets are used subconsciously as a sponge for the stresses and strains of our lifestyle.' This is very true of parrots, especially now that hand-reared young ones are so readily available. In advertisements, such birds are often described as 'cuddly tame'. They may attract a very different type of person from the one who would normally consider keeping a parrot as a pet (*see* Cuddly-tame.)

Formerly people kept parrots and other birds for one reason only: they enjoyed their companionship. The more intelligent species such as parrots know instinctively if a person, their owner or otherwise, feels warmly towards them, because that person loves, respects and admires birds. Alas, today, many parrots are kept for the wrong reason. Parrots, like the more prestigious reptiles such as iguanas and boa constrictors, are deemed to be fashionable and trendy. Other people keep parrots only for commercial purposes. Some 'owners' have little or no affinity with birds and animals and have not the slightest idea how to behave in their presence. When he or she shouts, gesticulates or moves large or strange objects near a parrot's cage or aviary, he or she is not thoughtful or sensitive enough to be aware of the fear being instilled in the parrot. Such a bird can live in a constant state of stress and tension. Little wonder, then, that such parrots are difficult to tame (if not hand-reared) and show no inclination to 'talk'. In the case of a sympathetic and observant person, who was tuned in to the psyche of a parrot, the same bird would be a happy, lively, loving companion.

Dr Rosina Sonnenschmidt is also very aware of how a parrot owner can unknowingly influence the psychological welfare of his or her bird. She wrote: 'If someone tells me about his "difficult" parrot I first have a conversation to find out what problems of the human psyche are being projected on to the bird. It is incredible how precisely our pets react to our thinking or rather, to our attitude. Is there a positive basic mental attitude or a negative one?

'Even if a bird owner argues, "It's not my fault, the bird was difficult when it came to me", we have to ask why some of us attract only difficult birds!'

A parrot of a sensitive species who lives with a person who is constantly under stress may suffer as a result. Pamela Clark wrote of Grey Parrots:

> ... our Greys clearly feel the 'energy' of our emotions. They pick up on, not only the quality of the emotion (anger, sadness, stress, joy, serenity and confidence) but also the direction of the emotion. If our expression and our tone do not match the 'felt sense' of the action the bird receives, confusion will result.
>
> For example, if we experience chronic stress, our Greys will pick up on this, despite what might be our best attempts to smile and be loving and cheerful. They will sense that 'something is wrong' and they will have no ability to put this into some kind of perspective. If a parrot in the wild 'picks

up' or senses this type of message from a flock member, it means that life-threatening danger may be a breath away.

The result may be a companion who evidences a constant sense of unease or edginess . . . I find the least optimal home for a Grey is one in which the 'flock' members who are bonded to the Grey suffer ongoing self-absorption, and offer little awareness of the needs of the Grey beyond those needed for physical survival and perhaps a token session of interaction each night. Greys need an emotional connection to us so badly that such a home can be considered to be an abusive home, even if the bird's physical needs are being met. Those in such a situation frequently manifest symptoms of dis-ease [unease], such as feather chewing, nervously picking at their feet, or biting (Clark, 2000).

That parrots can interpret some of our emotions there is no doubt. The owner of a Dusky Lory was having many problems with the bird because her husband disliked it and it had been banished to the upper part of the house. The unfortunate lory had become very aggressive, no doubt due to lack of attention and stimulation. Her owner told me that when the lory was angry with her, the lory would tell her: 'Shut up!' Clearly the lory had interpreted the feeling of anger directed at her when she was told to 'Shut up' and was then directing this anger back at her owner in different circumstances.

ENVIRONMENTAL ENRICHMENT

A term relating to captive animals that was seldom heard in the 20th century but is the new by-word in the 21st, is "environmental enrichment". In simple terms, this means ensuring that its living space is interesting and contains plenty of items to occupy it. These should give it the opportunity to take part in natural behaviours, such as climbing, swinging and gnawing. It is easy and inexpensive to make toys using lengths of knotted leather (knotted to prevent accidents) and small pieces of wood.

Parrots can even have the chance to solve puzzles to obtain food. Several manufacturers have created different containers, called feeding toys, from which parrots must manoeuvre food through holes or pull levers until the food falls out through the bottom. Sonny Stollenmaier filled a wicker picnic basket with items for his two Moluccan Cockatoos. It contained about 20 small toys and, on some days, food items such as a slice of whole-meal bread. Before they could obtain these, they had to learn to undo the latch on the basket, which was similar to a trouser belt.

Environmental enrichment can also involve placing animals in social groups. This can solve common problems of companion parrots, such as screaming and feather plucking. A prime example can be seen, ironically, in the largest parrot rescue centre in the UK. In one aviary there are more than 100 Grey Parrots. About three quarters of these birds arrived in a badly plucked condition yet the vast majority of them were perfectly feathered after a few months. The stimulation of being in a flock, the best possible environment for these highly social and non-aggressive parrots, was what was needed to solve a problem that is so difficult to cure in a cage environment.

Scientific studies in the 21st century have demonstrated that an enriched environment results in parrots that are healthier physically and mentally. In the USA a study was carried out using two groups of eight 16-week-old Orange-winged Amazon Parrots (*Amazona amazonica*). One group lived in the most boring surroundings possible, with only food and water and a couple of perches. After a year six of these birds began to shred or pluck their feathers. Examination by a vet confirmed that this was not due to an undiagnosed health problem. The condition was reversed after a few months by placing them in an enriched environment. The other group had to search for their food or to pull levers or chew through barriers to reach it. They had toys on which to chew, swing and climb in their environmentally enriched area. These birds remained in good plumage and did not shred or pluck their feathers. All 16 Amazons received the same healthy diet (Meehan *et al*, 2003, Meehan *et al*, 2004).

POSITIVE REINFORCEMENT

Positive reinforcement is the term used when an action by an animal is immediately followed by the trainer causing something good to happen. A parrot will soon connect these two. A behaviour is reinforced (maintained or increased) by adding something desirable to the environment as a consequence and when the parrot repeats this behaviour because it remembers the favourable consequence. For example, if you want your parrot to step up on to your hand and you praise him lavishly in a voice that indicates your pleasure, or you rub his head, he will want to step on to your hand. The desired action is *reinforced* with praise, head-rubbing or something else that is pleasurable. The reward must be considered desirable by the parrot or it will not have a reinforcing effect. The parrot should therefore be carefully observed to discover what it really likes. Positive reinforcement methods of training animals do not use intimidation and lead to a better and more trusting relationship between bird and owner than those methods whereby the owner forces his wishes on the bird.

Sonny Stollenmaier states: "Your parrot will always indulge in the behaviour that it considers the most reinforcing at any one time and will not perform a behaviour that it predicts (due to past experience) will be less rewarding. For example, if you lift your hand and offer it to your parrot (the antecedent), your bird steps up (the behaviour) and you reward it with a nut (the consequence), thus stepping up has been positively reinforced. Once learned, this behaviour will be offered again the next time you present your hand, as the parrot has learned that something good will follow. However, even the best-trained and most well-behaved parrot will sometimes refuse to step up. This can happen when, for example, the bird is playing with a toy and considers this to be more rewarding than the nut and so the behaviour of playing is maintained over the requested behaviour of stepping up. It is usually best not to push the point but to leave the parrot and to ask it to step up again a few moments later. However, if you offer something that your bird would find even more rewarding than the toy, it will abandon the toy and step up in order to receive the new item" (Stollenmaier, in Low, 2006).

The term positive reinforcement is self-explanatory but some terms used by animal behaviour consultants might be confusing. In the 1920s psychologist B.F.Skinner suggested that the best way to understand behaviour is to look at the causes of an action and its consequences. He called this approach operant conditioning. The term is often used by animal behaviourists.

As an example, good behaviour can be repeated and then becomes a habit if the incentive is strong enough. If your parrot refuses to enter its cage at a desired time, withhold a favourite treat until you want him to enter the cage. If you do this every time, your parrot will enter voluntarily because it knows that a piece of pomegranate (or whatever it relishes) is waiting for it.

RELATIONSHIPS BETWEEN PARROTS

The companion parrot owner, rather than the keeper of parrots in aviaries, may have given little thought to the complexities of the relationships which occur between parrots in a family or flock situation. This is, for me, one of the most interesting aspects of keeping parrots. Those who take the time to study this will uncover events and stories almost as absorbing as those in a TV soap opera. Because most parrots in aviaries are kept in pairs, rather than flocks, there is little opportunity for the average keeper to observe this fascinating side of parrots. But even family life will uncover some surprises.

As an example, a pair of Stella's Lorikeets reared their first youngster. It was a male. After he left the nest the male transferred his affection from the female to the young male. They spent hours wrestling together on the aviary floor. The female wanted to join in and would sometimes fly down beside them—but they would never let her play with them. To take an anthropomorphic view, this might be described as 'men behaving badly'. When resting, the two males would sit together, preening each other. The female was always excluded, and had to feed last, despite the presence of more than one feeding dish. Sometimes when the males were excited one would display to the other. (However, this should not be interpreted as homosexual behaviour.)

When watching the interaction between parrots, it is so easy to draw parallels from human life. While this applies in the study of probably the majority of the more intelligent bird species, it is perhaps easier for the human observer to understand them in parrots. This is one of the reasons why parrots are such endearing companions. It is why cockatoos, macaws and Amazons, for example, are such popular pets. They wear their hearts on their sleeves.

She'll never get off the ground

SOCIALISATION

It is very important that a parrot be involved in as much activity within the family as possible. The humans in the house are members of its flock; parrots are highly social creatures which, in nature, interact with many other flock members, not only their mates. If the parrot is not properly socialised with members of the family other than the main care-giver, it may become overpossessive of that person. In fact, this might happen anyway, but an attempt to avoid this should be made from the outset. In any case, there are several advantages. In the absence of the main care-giver, the parrot can enjoy the same quality of life, being taken out of his cage and moved around the house instead, of, perhaps, being confined to the cage.

Assuming that a young hand-reared parrot is obtained, the ideal scenario would be for all members of the family (except young or untrustworthy children) to be able to handle it and to build up a relationship with it. In order for this to occur, each reliable family member needs to spend a few minutes with the parrot every day, preferably in a room away from its cage.

The location is important because, at first, in strange surroundings, most parrots will lack confidence. This strengthens the relationship with the person because he or she is familiar—and everything else in the room is unknown. Of course, after a while, the bird will become familiar with the room but by this time the friendship between parrot and human is growing.

The benefits are two-fold. If every trustworthy member of the family takes the parrot to their room for a few minutes daily (it must be certain that no teasing or abuse will occur), the day becomes much more interesting for the bird. Instead of seeing the same view, day in, day out, its horizons have expanded, albeit slightly, and a better balanced parrot will result. It will also react better in unfamiliar surroundings because it is not used to being confined to a single spot. Each room might contain a perch in a convenient place so that the person does not necessarily have to be holding the parrot but might be getting on with some chore.

GENDER

The sex of a parrot does exert an influence on its personality and behaviour. Whether it is the male or the female which is the most gentle and affectionate depends on the species. It is generally accepted, for example, that male Budgerigars make better pets than females—and that was certainly my experience in my teenage years. Females were more independent and less affectionate. In strongly female-dominant parrots such as some lovebird species and Eclectus Parrots, the same is true. In Greys gender has little influence on the pet potential of a bird except that males can be more difficult for short periods when hormones are raging. In Amazons and *Pionus* Parrots I believe that females are gentler. In some lories, especially the larger *Chalcopsitta* species such as the Black (*C.atra*), it has been my experience that females are outstandingly more affectionate to people (when tame) than are males.

ORIGIN

These days no legally wild-caught parrots are imported in the UK. None at all enter the USA under normal circumstances—but there is an exception for small numbers imported by breeders' consortiums or for single pet birds going into quarantine when their owners take up residence in the USA. These are rare and would not involve birds for resale for the pet trade. Nevertheless, a few wild-caught parrots are still illegally smuggled across the Mexican border. (Macaws and Amazons would be the most likely species.)

In Europe, legislation to ban or restrict the importation of certain parrot species has caused a decline in the numbers of wild-caught birds available. For example, in 1988, according to official figures, 43,132 parrots were imported into the UK, and nearly all of these would have been wild-caught. By 1997 the official number of imported parrots was 5,500, including 302 Grey Parrots. Unfortunately, this number does not tell the whole story. Recent legislation permits the importation of birds from Belgium, without quarantine, provided that they are accompanied by a health certificate signed by a vet. (Normally all parrots which enter the UK must be quarantined at Ministry of Agriculture-approved premises for 35 days.) Inevitably, the Belgium-exception system has been abused. How many of the numerous wild-caught Greys offered for sale in 1998 had arrived via Belgium (possibly with a health certificate) and how many had come through the usual official channels is unknown.

The importation of wild-caught parrots and other birds into the UK and other EC countries did not cease until October 2005 when there was a temporary ban because of the fear of an outbreak of the H5N1 virus of avian influenza. The temporary ban was in place until June 30 2007 when it was replaced by a permanent ban. It is likely that illegal smuggling of parrots into the EC will occur from time to time thus it is still important that buyers can distinguish between captive-bred and wild-caught birds, because the origin of the parrot will have an immense impact on its behaviour, health, welfare and longevity. Sadly, a large proportion of wild-caught parrots die within a few weeks of importation. When many wild-caught birds, under terrible stress from the trauma they have endured, are crammed together for export and shipping, disease proliferates.

A wise buyer will take the trouble of locating a breeder, rather than buy a bird of unknown origin at a show. Responsible breeders put closed rings on their young birds. These rings usually show the breeder's initials and the year of hatching.

Diseases such as salmonellosis and aspergillosis were common in recently wild-caught parrots. (Aspergillosis is a respiratory disease; in some cases the birds can be seen and heard wheezing—*see* also Wheezing). Both diseases are normally fatal. Although many affected parrots looked healthy they were probably terminally ill when purchased and would survive only a few weeks. As most people do not take the precaution of quarantining new arrivals away from the other birds they own, there is a danger that infectious diseases like chlamydiosis (psittacosis) and salmonellosis will be passed on to other birds, with fatal consequences.

People who inherit or take on a wild-caught parrot should remember that it can take several years for the terrible memories of capture and export to fade. Perhaps for some birds they never do. Some wild-caught birds are terrified of sticks, for example, or anything that resembles them—perhaps long-handled brooms. Others remain frightened of men all their lives. We shall never know what they have suffered. I can only ask anyone in possession of a wild-caught parrot, especially a Grey, never to forget what the bird has experienced. If, with love and patience you do gain its trust, be ever-aware that a careless or unsympathetic action could revive memories that spell fear into its heart, and perhaps trigger feather-plucking or another nervous habit.

In the UK many Orange-winged Amazons were imported from Guyana. Because the price was comparatively low for a parrot, these birds were bought by people who could not afford the price (more than double) for a hand-reared Grey. In an article in the magazine *Just Parrots* a reader described how she had obtained such an Amazon in a garden centre. In this case the bird, called Gunner, was unlikely to have been cheap, because she had had a previous home and was said (inaccurately) to talk. She was purchased out of pity for her plight. She shook with fear every time someone approached her.

Gunner's new owner was well aware that she would need time and patience to win her trust. After one month she had not been persuaded to leave her cage and was aggressive inside it. During the second month she took food from the hand but refused to leave her cage. The answer came in the form of a new large cage with a top opening. She came out—and it was possible to train her to step on to a perch.

That took only one week. After six months she came out of the cage on her own and allowed herself to be carried around the room on the perch. By this means she was taken out and put back into the cage. But she still growled at everyone except her rescuer.

After one year came a breakthrough. Gunner allowed her owner to stroke her head. After that, every time her owner passed the cage, she lowered her head to be tickled. One day when her owner was eating some cake, she flew to her (for the first time ever), helped herself to some cake and flew back to her cage with it. This became a habit, so a stand was placed by her owner's chair and the Amazon was soon flying backwards and forwards.

After three years she had stopped growling at women but still disliked men. She was terrified of strangers. After four years her owner felt she had gained 90 per cent of her trust—but Gunner still declined to climb on to her arm. Her owner wrote: 'What kind of hell did this poor little parrot go through, being caught and taken from the wild? What kind of hell would make her fear someone who has loved her for the past four years?' (Sawkins, 1997).

READING PARROT BEHAVIOUR

As mentioned above, we can learn to interpret parrot behaviour by studying individual birds. Unless we can do this, we cannot even begin to understand why parrots behave as they do. The owner of a companion parrot must be able to recognise nuances of behaviour that indicate the bird is relaxed in his or her company. If these behaviours do not occur, training and teaching to talk will be difficult to achieve at that time. Some pointers to a relaxed bird are as follows:

- Preening the tail. At such moments a parrot or other bird is most vulnerable to attack from a predator since it cannot be on the alert when its beak is touching its tail. If a parrot tail-preens in your presence, it is a clear signal that you are not regarded as a threat.

- Simultaneously fluffing and ruffling all the contour feathers—clearly audible in larger parrots. It is almost used as a greeting to a greatly trusted human companion. For example, every morning when the owner enters the room, a parrot is likely to rouse its feathers in this way. It is not repeated and only lasts for a couple of seconds. In an aviary bird, this behaviour can be a sign of contentment at reaching the spot in the aviary where it feels most secure. Sometimes the action is concluded by a rapid side to side movement of the tail.
- Stretching the wing or wings. A parrot which feels threatened is very alert and ready to fly at a moment's notice. It therefore stretches its wings only when it feels totally secure.
- Beak grinding—usually seen when the bird is settling down to sleep.

2
CAUSES OF UNACCEPTABLE BEHAVIOUR

In the previous chapter factors which affect a parrot's behaviour were examined. But some behaviour problems are often difficult to fathom. This is especially true in the case of feather plucking. There are so many possible causes. In order to try to establish the cause of any behavioural problem you need to:

1. Try to put yourself in the parrot's situation—look at the world as the parrot sees it.
2. Be aware that if the behaviour of your parrot suddenly changes, there is a very good reason for this. It could be hormonal, but on the other hand some external influence may be to blame. Ask yourself the following questions:

- Has anything changed in the normal procedure of the household? Or in the human members? The absence of a person to whom the parrot is attached can profoundly affect it; likewise, the addition of a person who is unsympathetic or even abusive.

Case history
Pionus are a genus of neotropical parrots with short tails and scarlet under tail coverts. They are quite often kept as pets. A member of the *Pionus* Breeders' Association told a disturbing story (Spring 1998 newsletter). She had a Blue-headed *Pionus*, described as 'beautiful and the love of my life'. A little over one year after she obtained her, the *Pionus* started screaming incessantly. Her neighbours complained. She sold the *Pionus*. Later

she learned that her boyfriend had been abusing her bird. The form of abuse was not specified. If a human child, too young to speak, was being abused, it would probably react in the same way—by screaming. Sadly, the 'love of her life' was not an enduring love. Its strength was not great enough for her to make an effort to discover the cause of the Pionus' disturbed behaviour. The first thing she should have asked herself was: 'What has changed?' Even though she could not have known about her boyfriend's conduct, she might have connected the changed behaviour with his presence.

- Next question: has anything changed in the cage or its immediate environment? Many parrot owners have no idea how sensitive their birds might be to the sudden presence of strange objects. This is why new toys or fresh-cut branches should be placed near the cage for a few hours before being placed inside.

Case history

A local parrot rescue centre received a telephone call from a couple who could no longer tolerate the screams of their parrot, which had suddenly become very noisy. If necessary, the rescue centre would have taken the parrot, species unknown by the owners. The answers to a number of questions revealed the bird's identity as a Patagonian Conure (*Cyanoliseus p.patagonus*). This is a charming, potentially very affectionate, long-tailed parrot, the largest of the conures. It would be kept as a pet more often if its voice was less powerful. Carol, who runs the rescue centre, went through the whole range of questions designed to discover the reason for a change in behaviour. One of them was: 'Have you altered your parrot's cage in any way recently?'

On reflection the owner said: 'Yes, we have added a swing.' It was at about that time that the conure became noisy. It also started to pluck itself. Carol suggested that the swing should be removed from the cage. Several weeks later she received a telephone call from the owners. They had removed the swing and immediately the conure had quietened down. It had stopped screaming and it had stopped plucking. They were delighted because they had not wanted to find a new home for it. So many

problems of this kind could be solved if only the owners made more effort to discover the cause.

- Is some element of your parrot's diet having a detrimental effect on its behaviour and well-being? It is not generally realised that parrots can suffer food allergies, just as humans can.

Case history

While at a convention in Canada in 1998, I met the owner of an Eclectus parrot who related a very interesting story. Eclectus are large parrots; the males are emerald green and many make superb pets and wonderful talkers. Stefanja had hand-reared her Eclectus with loving care from the age of six weeks. He had a large cage, many toys and at least three hours out of the cage daily with his caring owner. He had a bath three times a week and his diet was exceptionally varied. It sounded like a recipe for a happy and healthy parrot. But he was a sick bird, with black stress marks on his new feathers. His behaviour was becoming increasingly aggressive and unpredictable. Stefanja described to me his screaming sessions which sounded like 'the equivalent of banging his head against a wall'.

He had been under veterinary treatment since he was six months old when, in addition to the bouts of sneezing which had always plagued him, he had a nasal discharge. The vet carried out blood tests and various cultures. The blood tests were normal and the choanal cultures did not reveal a problem there. (The choana is a passage which leads from the nostrils.) With treatment the sneezing decreased but did not stop. The Eclectus, called Zebedee, had itchy and flaky skin. The vet did not know what was wrong with him.

A month after the tests Zebedee started to have mild seizures. The vet suggested the possibility of epilepsy or a brain tumour. A head X-ray was recommended and a course of valium. Stefanja refused to put him through any more tests or treatments and Zebedee developed pneumonia. In addition to swollen eyes and nasal discharge, he had an allergic reaction to the oral and injectable antibiotic that the vet was using. At this point it occurred to Stefanja that Zebedee might have an allergy. She suggested this to her vet who said the Eclectus might be allergic to the dust from her Cockatiel. Stefanja thought that this was unlikely.

One day a friend mentioned that her child was not allowed to take anything containing peanut butter or other peanut products to her day care centre because one child had a life-threatening peanut allergy. Stefanja realised that if her Eclectus had a similar allergy it would explain why he woke up with minimal symptoms, but that those intensified during the day. By the evening he was grumpy, aggressive and screaming. She decided to eliminate all forms of peanut from his diet. First to go was the peanut butter on toast. She checked the Tropican pellets and his Nutri-berries (a manufactured, textured and nutritionally balanced food) and found that both contained peanuts. Most seed mixtures contained either peanuts or pellets or crumbles which had peanuts in them. A special order was made to a company for a peanut-free food. Within 24 hours of feeding this and withdrawing all items containing peanuts there was a dramatic improvement in Zebedee's condition.

A week later Stefanja tested her theory by giving him one Nutri-berry and some peanut butter. Both gave him almost immediate sneezing fits and nasal discharge. The source of the problem had been proved. On another occasion, Zebedee was given a little Vietnamese food which appeared to be peanut-free. Immediately, the old symptoms recurred. When Stefanja contacted the restaurant which had prepared the food, she was told that peanut oil was used for cooking. By this time seizures were occurring which lasted about 24 hours. During this period Zebedee was kept warm and his cage was covered.

By then Stefanja had learned a lot about foods which might contain peanuts. She was told, for example, that other types of nuts may contain traces which are acquired during the grinding process. Zebedee's diet now has an emphasis on fresh foods—organic when possible. His toast is moistened with organic flax seed oil and sprinkled with a vitamin supplement. He enjoys various green foods, such as dark green lettuce and dandelion, sometimes with the root attached. He also eats sprouted and dry seeds, apple and willow twigs. His 'supper' consists of cooked brown rice with a cooked vegetable. This is varied with goat's milk, yoghurt, tofu, hard-boiled egg or well-cooked meat. Spirulina is sprinkled on this meal.

In this way Zebedee became a happy and healthy bird again. His stress-marked feathers were moulted and replaced with perfect ones. His plumage shows a wonderful iridescence. If only there were more parrot owners like Stefanja who was determined to get to the bottom of a very serious problem. . .

PAST HISTORY

If your parrot had a previous home or homes, try to find out as much as possible about his history. If he was obtained from a breeder or dealer when very young, how old was he? Some breeders provide a hatch certificate which shows the bird's date of hatching, parentage, name of the breeder, etc. The most important information is the date of hatching. The psychological problems suffered by so many hand-reared parrots, especially cockatoos, stem from the fact that they were force-weaned. Many people purchase hand-reared parrots through a dealer or pet shop. The hatch date is unknown. They might be told that the parrot is weaned; whether or not this is true is something they will not know until a day or so after the parrot has been in their care. If the bird is older, it may be impossible to obtain details which would assist you in understanding his behaviour.

Case history

A Blue-fronted Amazon Parrot, obtained at the age of eight years, was very noisy and his screaming was a problem. He had the freedom of the house and company of people for most of the day—more stimulation than many parrots receive. When I asked the owner about his previous history, she told me that he had previously been kept in a caravan and never let out. In these unstimulating circumstances, it may be that screaming had become a habit which was hard to break. However, this behaviour was greatly reduced by the provision of gnawable toys which kept him quiet for long periods.

• All aspects of the parrot's day should be considered in detail. This reduces the likelihood of missing a vital clue. Again, try to think like a parrot. If you only consider the problem from your point of view, you will never solve it.

HORMONAL INFLUENCES

At the onset of sexual maturity and for two or three months every year, the behaviour of some parrots—especially males— may change dramatically. Hormonal influences cause this.

Kristin Shay wrote of her Grey Parrot, Moshi:

For the past two years [since he was two and a half] there has been a danger period starting at the end of January and lasting for about four weeks. Moshi Jekyll becomes Moshi Hyde during this time and has to be approached with caution, especially around his cage and for the first hour of the morning. Normally he's only too anxious to leave his cage and step out of it on to my hand with alacrity, as well as repeatedly calling 'Want to come out'. But the reverse is true during this period and eventually the need to void his system [he keeps his cage clean] drives him out and we more or less carry on as usual for the rest of the day. The return journey to his cage at night has to be equally carefully handled and, once safely installed there, he will lunge at me with intent, as I quietly close the cage door and draw the blanket cover together.

The last behaviour is totally out of keeping since he is normally extremely affectionate at night. After the hormonal upsurge has run its course, Moshi reverts completely to his old self, becoming even more affectionate except for no longer welcoming caresses when in his cage (Shay, 1999).

ILL HEALTH

If you have tried to deduce the reason for a change in behaviour and you have failed, the next step is to consult an avian vet. These members of the veterinary profession are listed in some bird magazines such as (in the UK) *Parrots Magazine.* A vet in a small animal practice which sees few bird patients may not be aware of the wide range of tests which can be carried out now to diagnose disease. Ill health can have a profound effect on the personality and behaviour of a parrot.

ATTENTION-SEEKING

Unacceptable behaviour is often caused by the parrot's desire to grab your attention, especially if it feels neglected. Sometimes, over the years, an owner can start to take a parrot for granted and not give it so much attention as it formerly had. As it is not receiving positive attention (praise and rewards for behaving well) it will scream or fling seed. Before you get annoyed with your parrot, please ask yourself whether this wonderful companion, who has so few ways of amusing himself, is receiving all the attention he deserves from you...

On the other hand, new owners should be very aware of the dangers of over-indulging a new parrot—especially a young cockatoo or macaw who will soak up all the affection and attention lavished on it. This can be overdone. You can unwittingly be training it to expect so much of your time that you become a slave to it. When it is no longer possible to maintain this level of attention, screaming and/or feather plucking will result. A young parrot must learn right from the outset that it must amuse itself for much of the day. It must not be totally reliant on you or other family members. When this is allowed to happen, if family circumstances change and perhaps only one person remains to care for the bird, the demands it makes will be so extreme that that person cannot cope. The unfortunate parrot will be sold or given away.

THE QUICK FIX DOESN'T WORK

Many parrot owners are unable to correct undesirable behaviour in their parrots. This is usually because they do not persevere long enough. Sometimes it is because they have been given very poor advice. The fact is, there are many conflicting ideas, often due to inexperience on the part of the person offering the advice. This might be another pet owner, a self-styled behaviourist (professional or otherwise) or even a veterinarian. Generally speaking, vets do not have enough knowledge concerning the behaviour of companion parrots to give sound advice. When asked they give their opinion—but it might be far wide of the mark. The people most qualified to give advice are not necessarily professional people. The two main requirements are a deep knowledge of parrots and their behaviour, and a sympathetic attitude.

If someone gives you advice on handling or training a parrot which you instinctively feel is wrong, ignore it. If it is wrong, it could do irreparable damage to the relationship between you and your parrot. An example of this was related in a magazine. The advice was unbelievably bad—but as it was given by a trainer who had developed his own bird show, the owner of a Moluccan Cockatoo decided to follow the advice. What she perceived as a problem was that the Moluccan Cockatoo would not stay on its stand. It was unrealistic to expect this, anyway. She was told by the trainer that every time the cockatoo got off his stand, she or her husband were to pick him up *at the base of the tail* and place him back on his stand. To handle any bird in this manner would be appalling, in my view. The husband, to whom the cockatoo was bonded, followed the advice. Almost immediately, the cockatoo would have nothing more to do with him. He would turn his back and refuse to even look at him. For 18 months the cockatoo acted as though the husband did not exist. How could the unfortunate bird possibly understand why the man was behaving in this way? (*see* Punishment). For two years the cockatoo had acted in the same manner without being corrected. Presumably it previously had known only sympathetic handling. Suddenly to be assaulted in this way must have been a frightening and puzzling experience. Surely common sense should have told the man that? No caring person could have brought themselves to follow such insensitive advice. If they cannot behave with dignity towards their parrot, they do not deserve the privilege of looking after it.

In the UK, several series of television programmes have focussed on undesirable behaviour in pets and how to reverse the problem behaviour. Usually they show an animal behaviourist giving advice, and returning a month or so later to a reformed animal. While perhaps some bad habits in dogs and cats might be reversed in such a short period, this is much less likely to be the case with parrots. These programmes inspire people with unrealistic expectations. In the few instances where parrots have featured, the inference has been that the techniques suggested have been successful when, in fact, no rapid results should be expected in that particular time scale or the problem has been greatly simplified. Or even misunderstood.

In a bird which has never been tamed or which has been mistreated, it could be one to two years before significant improvements are seen. Most people give up long before this stage is reached. In some cases a structured training programme is the wrong approach and little will be gained before the bird's trust has been won. It may seem like a 'Catch 22' situation. Many people have neither the time nor the patience to deal with a parrot which has severe behavioural problems. They might have taken on such a bird because it was cheap or free, with little or no understanding of the difficulties involved in rehabilitating it emotionally. In unsympathetic hands its behaviour will worsen—until it is given away yet again. Believe me, a free parrot of a species which normally has a fair monetary value is seldom a bargain.

A parrot's early experiences can (as in a human child) have a big influence on its subsequent behaviour. This is especially relevant in the case of a parrot which is re-homed. Someone who takes on one which has had one or more previous homes should throw out of the window all the theories relating to what parrots do or what they like. Don't make assumptions! Give the parrot choices! Only in this way will discoveries be made which help it to feel more at home in its new environment.

Behavioural problems are never easy to solve. There are no set guidelines which can be followed. The precise circumstances which cause a parrot to start plucking, for example, are hardly ever duplicated. The owner of a plucking bird could carry out all the suggestions under Feather Plucking which will help a bird to break the habit and to become healthier—but, unless the root cause is addressed, the outcome will not be successful.

It is also unwise to generalise concerning parrot behaviour or training. The suggestions made in this book are just that—not hard and fast rules. Parrots are individuals—and what works for one will not be effective with another. There are a multitude of variables to take into account. These include the personality and knowledge of the owner, whether the bird is young or older and untrained, whether it is hand-reared, parent-reared or wild-caught, and if it is in good health. Trust between parrot and owner is often a key factor.

The species most often thought of as companion parrots are Grey Parrots and Amazons, macaws, conures and cockatoos. But there are a number of smaller species, including Budgerigars, Cockatiels and lovebirds, which are kept in thousands. In some respects their behaviour is very different to that of the larger parrots. Their capacity to learn may be less. I am not referring to mimicry, which is often outstanding in male Budgerigars and Cockatiels, but to other types of training. Methods described for the larger parrots may be totally inappropriate and ineffective.

Part Two

WHAT IS A PARROT?

3

THEIR LIVES IN THE WILD

The word 'parrot' can be used to denote any bird within the order Psittaciformes. Within this order there are two families, the parrots (Psittacidae) and the cockatoos (Cacatuidae). There are more than 300 species, many of which do not include the word in their name: for example, macaw, cockatoo, lovebird, lory and conure. The word parrot can be used to relate to the 'typical' parrot. These are medium-sized parrots with relatively short, broad tails, such as Greys and Amazons. Parrakeets are long-tailed parrots, and those from the neotropics are called conures. The name parrakeet is not used in Australia.

The word psittacine (pronounced sit-a-seen) is an adjective; it is often used incorrectly as a noun, especially in the USA. This is the equivalent of calling a pig a porcine! Worse still, the term 'hookbill' is sometimes used instead of parrot. This is confusing as it could apply equally to birds of prey. So what distinguishes the parrots from all other birds? The main features are the curved bill, exposed fleshy cere (bare skin above the beak)—both of which features are shared by birds of prey—usually bright colours and the arrangement of the toes. All members of the parrot family have two toes pointing forwards and two backwards; this form is known as zygodactylous. Virtually all parrots have some bright colour in the plumage—green, red, yellow or blue—exceptions being the two species of Vasa Parrots from Madagascar. They are grey or brownish and, with their curved bill, might even be mistaken for a small bird of prey, because some of the latter can turn the outer forward toe backwards or forwards so that two toes point in each direction. When the exceptions are included, it might seem that the parrots are not such a well defined group. But in fact they are, and few members of the parrot family could be mistaken for anything else, because along with the easily definable characteristics there are others, such as the way they move and their often inquisitive demeanour and usually sociable habits.

Parrots occur naturally in the tropics or subtropics. There is a tendency to think of them as rainforest dwellers. Many species, especially in South America, New Guinea and Indonesia, occur in this type of habitat. The majority of parrots originate from forested areas. Others are found in mangroves and swamps (Grey Parrots), coconut palms (lorikeets), seasonally flooded areas such as the vast Pantanal (Hyacinthine Macaw), sparsely timbered grasslands and coastal dunes (the Blue-winged Parrot in Australia), savannah woodlands, open grasslands and parks (Galahs), and in gardens, plantations and wooded country (Ceylon Hanging Parrot).

We associate parrots with swaying palm trees and tropical forests, yet they also occur in the subantarctic—literally alongside penguins. The Antipodes Green Parrakeet is a large, all-green version of the Red-fronted Kakariki, which is a much loved aviary bird. The bleak Antipodes, a far outpost of New Zealand, measure only eight square miles, are mainly covered in tussockgrass and are continually buffeted by strong, howling winds. Many parrot species are not strangers to snow and frost. Even in Australia, near the Great Dividing Range, lorikeets, parrakeets and cockatoos live among snow for brief periods. And the Kea, New Zealand's fascinating mountain parrot, is now mainly confined to high alpine reaches; agriculture has encroached upon its lowland habitats.

No parrot lives in colder climes than the Austral Conure of Tierra del Fuego at the southern tip of South America. In Michael Andrews' classic book *The Flight of the Condor*, he recorded how he suddenly saw 'two pairs of the Austral parakeet sitting on a stump right in front of us. To see parakeets next to glaciers is as odd as to find penguins in the desert. . .'

FEEDING HABITS

The fact that parrots occupy such a wide range of habitats indicates that their food sources are equally diverse. Another misconception relating to parrots is that they are primarily seed eaters. The fact is that the first members of the parrot family to be domesticated were Budgerigars, certain other Australian parrakeets, Cockatiels and lovebirds, which can exist mainly on seed, as this forms such an important part of their diet in the wild. They adapted well to captivity because seed (albeit hard, whereas most parrots feed on tender, green seeds) was easily obtainable and required no preparation. The habit of treating all parrots as seed eaters was to prove disastrous for some other species (such as lorikeets) in early attempts to keep them in captivity.

A number of parrots are specialist feeders: lories and lorikeets feed primarily on pollen and nectar, some large macaws exist on oily, fibrous palm nuts, and New Guinea's red and black Pesquet's Parrot is a fig-eater, as are two groups of small fig parrots from Australia and New Guinea. The Glossy Cockatoo from south-eastern Australia feeds mainly on casuarina nuts and another member of the *Calyptorhynchus* genus, the Yellow-tailed Black Cockatoo, has developed an elaborate technique for extracting wood-boring grubs from trees. In South Africa, the Cape Parrot feeds predominantly on the fruits of the yellowwood trees (*Podocarpus*).

Some species are sedentary, living all their lives within an area of a few square kilometres. Others are nomadic. If they have specialised food requirements, they may have to search over a wide area, perhaps moving long distances at certain times of the year in order to find their food. Lories and lorikeets fall into this category. In some areas there may be no flowering trees for considerable periods. Lories have to be adventurous (willing to travel), inquisitive and learn to take advantage quickly of new food sources, if they are to survive.

To us, these birds often appear very intelligent because they are so quick to learn and so inquisitive. What they are actually demonstrating is not intelligence but a readiness to adapt these natural traits in a captive situation. Let me give an example. Some of my lorikeets (nectar-feeding parrots) are kept in outdoor aviaries with an indoor section in an enclosed building. During the warmer months of the year they are allowed to sleep in the outside flights in their nest-boxes. During the colder months I shut them inside at night. One pair of Stella's Lorikeets objected to being shut in, so I allowed them to sleep out in the nest-boxes. With December came freezing temperatures and I persuaded the Stella's to go in at night because, if their hatch was left open, it let too much cold air into the building. They tolerated being shut in for two nights. Come the third night they refused. I could pick up the male and put him inside but while I was trying to persuade the female to enter he would come out again. I gave up! There are lights in the building, thus extending the hours during which the lories can feed on long winter nights. Just before turning off the lights I would, on the coldest nights, make fresh warm nectar so that each bird could fill its crop before the lights were dimmed. During the two nights on which the Stella's were shut inside, they must have realised that this was the advantage of not being allowed to sleep out of doors in their nest-box. So on the third night when they heard me enter the building to dim the lights, they left their nest-box in the dark, and entered the building for their share of warm nectar. And so I was able to shut them inside for the night.

By behaving like this they were demonstrating lorikeets' adaptability and the speed with which they can take advantage of a newly found source of food. Grey Parrots, for example, are highly intelligent birds, but I cannot envisage a situation where they would leave a nest-box in the dark in order to feed. Such behaviour would not be an adaptation of any form of natural behaviour.

Most parrots are, of course, reliant on trees for their food. In Australia, a few species of parrakeet feed primarily on the ground on the seeds of grasses. These include the Ground Parrot which, as its name suggests, does not live in trees. The other seed-eaters—the well-known Cockatiel, the Mallee Ringneck, Redrump, Hooded Parrakeet and the little Grass Parrots of the genus *Neophema*—only feed on the ground. They include parrakeets which are great favourites as aviary birds: Turquoisines and Splendids. Perhaps the most beautiful of all in this group, the Paradise Parakeet, is almost certainly extinct. It has not been seen in the wild since 1927. Its reliance on feeding on the ground was a death sentence, once the limited area in which it bred was taken over by cattle. Overgrazing and burning of grasses with their life-giving seeds to provide grazing for cattle left those birds with nothing to feed on.

DAILY ACTIVITIES

Even if they feed on the ground, most parrots rely on trees as roosting sites. Parrots are very much creatures of habit with a set daily routine. Most of them have special roosting sites where they spend the night, and some of these roosts are communal. Large noisy flocks gather at dusk, flying around and shrieking before they settle down. In contrast, some breeding pairs will roost in their nests. This is important throughout the year because nest holes are at a premium and would be taken over by other birds if they did not proclaim their ownership. There would never be enough holes for nonbreeding birds to be able to roost so securely—but there is safety in numbers, which is why so many congregate. At night they are more vulnerable to predators. The tiny little hanging parrots roost hanging upside down from a branch. In this way they are camouflaged by leaves. Were they to roost upright, they would be more easily visible.

When morning comes, just before or just after first light, parrots leave their roosting sites to visit their feeding grounds. Sometimes they must fly quite long distances. But even if their food source is nearby, this is the time of day of greatest noise and activity. Captive parrots are generally at their noisiest and most restless early in the morning. If possible they should be let out of the cage for a while then, even if only for a short time. They need to expend some energy at this time.

This is most noticeable in young hand-reared parrots, just before and after weaning. Some hand-feeders notice that their charges are reluctant to feed early in the morning and assume they are not hungry—and perhaps cut out a feed. Often young parrots are hungry but their need to flap their wings is so overwhelming they will not feed until they have done so. In no species is this more obvious than the large macaws. They may even be too young to fly but vigorously flapping their wings is an activity of prime importance. Most people do not have a lot of time to spare in the morning if they are preparing for work.

Nevertheless, they should try to find a way to include their parrot in the activities, if only for a few minutes. They might, for example, allow the bird out while they are preparing and eating breakfast. In fact they might get up a few minutes earlier to ensure that their parrot has some exercise out of the cage. It will be much easier for the bird to tolerate a number of hours confined inside after an early morning adventure.

In the wild parrots feed for two or three hours in tropical climates; then they rest because the sun is too hot for much activity. The hours around mid-day are the quietest ones. This is not necessarily the case for parrots which come from mountainous areas or cooler climates. They might be active for much of the day, seeking food, or they might venture out between showers. All species have another peak of activity in the late afternoon, when they feed and congregate with other members of their flock. Then they fly back to their roosting site. In flight, pairs keep together within the flock. A male and female Amazon or macaw, for example, fly almost wing tip to wing tip. Or family units—three or four birds—are identifiable for the same reason.

Parrots are exceedingly alert birds, always on the look-out for danger, even when feeding. In fact, some species which feed in flocks, notably ground-feeders which are especially vulnerable, post sentinel birds to raise the alarm if necessary. This behaviour is best known in cockatoos but may occur in many other species. Threats come most often from the sky in the form of birds of prey. This is why captive parrots are easily unnerved by unknown objects moving above them. This might be anything from a feather duster on a long handle to the silhouette of a large bird, even a gull, high above in the sky. Sometimes my Amazon, located near a window indoors, would cock her head, make a small sound which expressed concern, and continue to gaze skywards. Sometimes I could see nothing so I was certain her vision was better than mine. On other occasions the source of the fear, such as a Sparrowhawk, was evident.

Birds must be ever vigilant; those which are captive-bred do not lose the instinct to be watchful as it is a fundamental part of their existence. The unwary become prey. Even the largest of all parrots, the Hyacinthine Macaw, occasionally forms part of the diet of the huge Harpy Eagle. Large reptiles, such as (in

Australia), goannas, and feral cats and introduced mammals such as stoats also prey on parrots.

Field studies of parrots appear to indicate that they can relay information to each other. The feeding behaviour of the endangered Lear's Macaw (*Anodorhynchus leari*) has, in recent years, been studied extensively in Brazil. Biologists have frequently seen a flock fly into a potential feeding area—that is, one where licuri palms are present. Two birds from the flock will fly to inspect the feeding area while other flock members rest on a high tree close to the palms. Then the whole flock flies to inspect the area and either starts to feed or flies off. Perhaps communication regarding the ripeness of the licuri fruits or the presence of something alarming involves the decision to stay or leave. Parrots are so inquisitive that even if the two lead birds found a reason to leave the area, it would be likely that the flock members would want to see for themselves—unless, of course, a predator was present, in which case the alarm calls of the lead birds would clearly indicate this.

The parrot's ability to communicate information should not be overlooked in a captive situation. If your parrot is behaving strangely, don't dismiss its action, but just stop and consider the idea that it is trying to communicate with you. This is especially likely if it is performing some action which is not usually seen but is designed to attract your attention.

BREEDING BEHAVIOUR

Most parrots are monogamous, that is, they have only one mate at any time. At least from the time of mating or, on a permanent basis (depending on the species), the male stays with the female and helps to rear the young. When the chicks are very young he is the sole provider of food; he feeds the female at the nest. As the chicks grow the female leaves the nest and is away for increasing periods in her search for food.

It is known that in at least a couple of species a number of males attend to the female at the nest. One of these is the Eclectus Parrot. Of any parrot, this species shows the most extreme form of sexual dimorphism, that is, male and female look totally different. The male is mainly a vibrant emerald green with an orange beak, and the female is mainly red and blue, or red, with a black beak. This is almost a matriarchal society, for the female is the most important member. An unknown number of males feed her at the nest. As soon as the young fledge she lays another two eggs—or perhaps she is replaced by another female. Most males show a lot of respect for her. There is no close pair bond and mutual preening seldom, if ever, occurs. We do not yet know much about the breeding biology of Eclectus in the wild but it is possible that a group of males permanently tend to the female at a particular nest, with the female perhaps changing several times in a year. What we do know is that that the temperament of Eclectus is very different to that of most true parrots: an important factor for the companion parrot keeper to bear in mind.

Most adult females have a somewhat irascible nature. There are exceptions, but few are sweet-tempered. I have hand-reared countless Eclectus and have seen how, from an early age, some females will lash out at the human hand, even though they have never met anything but kindness. Male Eclectus, on the other hand, can make the most wonderful pets. They do not usually like to have their head scratched or enjoy close physical contact as do cockatoos and macaws, but they have their own way of showing affection which is, of course, related to their own unique courtship behaviour. A courting male Eclectus will rap his beak against the female's while making an intriguing 'bonging' sound. If you have a close relationship with a male Eclectus, then you will be honoured to be on the receiving end of this behaviour—and your nose will replace the female's beak!

Like most other parrots, Eclectus seek out holes in trees as nest sites. Some interesting discoveries about this were made on the Indonesian island of Sumba, where in 1989 and 1992, Dr Martin Jones led a team of scientists from Manchester's Metropolitan University. Parrot nests are seldom easy to locate, but using local parrot catchers, in 1992 it was possible to locate 122 nests. More than 85 per cent were in tree cavities, usually at the site of dropped branches. Apart from one species (the Red-cheeked), which nested in dead trees or stumps, the nests were in enormous trees, which averaged 35 metres (115 feet) high. The nest trees were usually the largest trees in the area. The parrots preferred trees of one particular genus, *Tetrameles*. Although less than 60 per cent of the trees examined were of this type, they contained more than 60 per cent of parrot nests. Some individual trees contained up to five parrot nests and one had four active Eclectus nests in it. It seemed that Sumba's parrots were not territorial (Marsden, 1995).

Some parrot breeders make a fundamental mistake in housing pairs of certain species near each other. It can be counter-productive to have pairs of Amazons or Rosella (*Platycercus*) parrakeets in adjoining aviaries. Males are very competitive and may spend more time trying to harass their neighbours than courting their females! On the other hand, some species need the stimulus of the presence of their own kind. Budgerigars and the little *Brotogeris* parrakeets from South America are social nesters. In the wild, they are found in huge flocks. During the breeding season the members of the flock nest in close proximity. In Australia, a number of pairs of the little Blue-winged Parrakeets (parrots in Australian terminology) can be found nesting within a few metres of each other in a patch of woodland, and Sulphur-crested Cockatoos congregate to nest close to each other along the banks of the Murray river:—not in trees—in the cliffs.

Some macaws also breed in cliffs. In the southern part of Piaui state, in Brazil, Hyacinthine Macaws nest in caves in the sandstone cliffs overlooking the dry forests. These are partly excavated by the macaws. Although there were about 100 metres between the closest Hyacinthine nests, a pair of Green-winged Macaws and a pair of their larger blue relatives nested only three metres (ten feet) apart. Much further to the south, in Bolivia, sandstone cliffs are used by Green-winged Macaws and two other parrot species. At Caquiahuara in the Alto-Madidi National Park, Severe Macaws and the smaller White-eyed Conures nest in the cliffs which are 60–90 metres high. Some of the nest holes are as large as two metres (six feet) in extent, 25 centimetres (ten inches) high and 15 centimetres (six inches) wide.

One of the most interesting facts to emerge from the study of this site related to four pairs of Green-winged Macaws in one locality. One pair would defend the cavity nest sites of the three other pairs in their absence. Pairs other than the rightful owners sometimes try to take over nest sites—and battles can ensue.

When one understands that not all macaws nest in tree holes, it becomes clear why some pairs in aviaries show a desire to nest on the ground. If the floor is not solid, some try to dig into it. For pairs which show no interest in a nest-box hanging high in the aviary, a horizontal—rather than an upright—nest-box on the ground might encourage them to nest. In fact many pairs of large macaws show a preference for horizontal nest-boxes.

Many parrots—wild and captive—are at their most aggressive in the vicinity of their nests, especially where nest sites are at a premium: if a pair does not actively defend its nest, it will be taken over by the same or a different species. Nest sites with an entrance which is only just large enough for the species to enter are usually preferred. This prevents larger parrots from claiming ownership of the cavity. A pair of Cockatiels, nesting at Burakin in Western Australia, were frequently disturbed by a pair of Galahs. The cockatoos were interested in the site because the entrance was large enough for them to enter. In captivity, some parrots refuse to enter a nest-box because the entrance hole is too large. I can recall at least one occasion when a nest-box entrance was reduced in size and the female, who had never previously entered, was sitting on eggs inside, within a month.

REARING CHICKS

As already mentioned, in Eclectus the female is the sole feeder of the young. In the case of most parrots, the female or—in a few species—both parents feed the very young chicks. As the young grow they are fed by both parents in the nest. The length of time the young spend there varies: from four to five weeks in many Australian parrakeets and Cockatiels; up to about eight weeks in Amazons and *Pionus*; up to 12 or 13 weeks in the large macaws.

INDEPENDENCE OF YOUNG

When they fledge the young will become the sole responsibility of the male, in species which quickly nest again. In others they might be fed by both parents for varying periods which are related to the length of time the young spend in the nest. If it is a short time they quickly become independent. If it is a long time it might be months before they are independent. The cockatoos illustrate this fact very well. The Galah has the shortest time in the nest—only seven weeks. Within a month of leaving the nest they are starting to find food for themselves, but they are not independent until six or seven weeks after leaving the nest. As an agricultural pest, the Galah is one of the best studied of the cockatoos. It is known that newly fledged young assemble in a creche tree during the day, until all the young in the nest have fledged. Then the whole family moves off to an area where food and roost trees are available. The Galah may be the only cockatoo which does this. The habit is related to the fact that the clutch size averages four eggs, and varies between three and six, and that the interval between egg-laying is an average of just over two and a half days. Therefore, there can be an interval of a week or more between the fledging of the first and last young ones.

Fledged young, which are perfectly capable of feeding themsleves, can be persistent in worrying parents for food. An interesting story related by a bird keeper in Australia, Bob Branston, shows how at least one pair dealt with this situation. These Bare-eyed Cockatoos (Little Corellas) and their two young visited his garden. They fed on a heap of old seed from his aviaries. This went on for some time. One day when the two young were sorting through the seed, the parents flew off, unnoticed by their offspring. They never returned! The young cockatoos hung around for several days before finally realising that they were on their own. The parents had left them with an easy food supply—and moved on.

There is very little information from the wild on the breeding biology of the non-Australian cockatoos. Few nests, or none at all, have been studied of most Indonesian cockatoo. In captivity the young of the largest white cockatoo, the Moluccan, leave the nest at about 13 weeks. The young of captive Palm Cockatoos fledge after 11 weeks. It is known that in the wild the young stay with their parents until the following breeding season. I would expect that the same happens with the Moluccan and Umbrella Cockatoos, for example. In captivity young birds will solicit feeding from a parent long after they are independent—and sometimes they are fed. They do not need the food; they do need parental attention.

In the large white cockatoos the clutch consists of two eggs. In the wild it may be that only one chick is reared. After fledging and, probably for many months, this young bird has the full attention of its parents. Is it any wonder that hand-reared cockatoos such as Umbrellas and Moluccans suffer such severe psychological problems when they are force-weaned at 14 or 15 weeks? I have never tried to complete the weaning of these cockatoos until they were between five and six months old. This seemed to me to be the natural weaning age. By weaning them too early, breeders are creating generations of problem birds, many of which are destined to die young or to end up in rescue centres. Most of them are also useless for breeding purposes, as they do not have the chance to socialise with their own species when they are young. This appears to be a much more serious problem with cockatoos than with macaws.

This is interesting because, in the wild, the young of the large macaws remain for many months with their parents. In one population of Hyacinthine Macaws which was studied in Brazil, it was noted that the young stay at least a year with their parents. They are still there at the start of the next breeding season. In Bolivia, in the Alto-Madidi National Park, a single Green-winged Macaw hatched the previous season (probably near the end of the year) was still following its parents about in July. As the three perched together, the juvenile would bob its head to solicit food, although it was perfectly able to feed itself. Sometimes its relentless begging irritated the parent to the degree that the adult would bite it, lunge at it or chase it away. The young one would fly off, then return, landing not quite so close to the parent as previously. This behaviour suggests attention-seeking rather than hunger.

THE STRENGTH OF THE PAIR BOND

One of the most interesting aspects of watching parrots in the wild is to observe the strength of the pair bond and family groups. This is not usually readily apparent in most perching birds, partly because male and female are not constantly together, as in the case of most members of the parrot family.

Even within a flock of parrots, paired birds are easily distinguished. They sit close together, often preening each other, and they fly almost wing tip to wing tip.

If one member of a pair dies, or is trapped or killed, the other will almost certainly grieve, perhaps for a lengthy period. When one member of a pair of finches dies, however, it will rapidly find a new mate. In the wild, finches (for example) are not long-lived birds, compared with parrots. If they don't produce young in their first or second season, there is a possibility that they will die without leaving any offspring. On the other hand, the lifespan of the larger parrots is long. Many do not reproduce until they are perhaps five or more years old and their breeding life might stretch over several decades—if they live that long. During that period some large parrots change partners; others probably remain faithful for many years or until one member of the pair dies.

From my own experiences, I know that the pair bond between two compatible white cockatoos (of the genus *Cacatua*) is one of the strongest in the avian world. These birds show affection for their mate in a way which is very easy for the human observer to understand. That a cockatoo would not want to desert its partner, even in death, is no surprise to me. On the Indonesian island of Flores, a Lesser Sulphur-crested Cockatoo (*C.s.sulphurea*) was shot from a flock of birds which were raiding a crop. Its body was hung up—in a pointless attempt to discourage other cockatoos from feeding on the crop. The dead bird's mate returned to sit in silence close to the body of its partner. How sad that a human being acting in such ignorance of the intelligence and devotion of cockatoos should deprive of its mate a species which can show more faithfulness than many human beings.

This devotion to their partner is one of the attributes which makes parrots among the companion animals most loved by the human race. In return for the pleasure and affection which they give to us, humans should take a more responsible attitude towards them. In the captive context, this surely means total commitment. A parrot is a pet for life—not until you grow tired of it. If you cannot make that commitment, a short-lived creature like a guineapig is the pet for you.

4

PARROT SPECIES AND HOW THEY BEHAVE

Parrots can be divided into a number of groups, whose members have very different characteristics: macaws, cockatoos, lories, parrakeets and the true parrots. There are approximately 350 species of parrots. A species is a form which differs in appearance and behaviour from any other. Until recently what constituted a species was purely arbitrary. With the advent of DNA technology, for the first time, there is a positive method of distinguishing species. It seems likely that this will result in the number of defined species increasing. This is why one can give only an approximate figure at the present time.

Currently, 84 genera of parrots are recognised. A genus is a group of species which are closely related, with each species having a different geographical range. The various genera have very different habits and behaviours. Without a basic knowledge of these, it is often very difficult to interpret parrot behaviour. There are several other points which should be considered, especially by the first-time parrot owner. He or she may have much experience with domesticated animals—but this is no qualification for life with a parrot. Parrots are very different from dogs, for example, all breeds of which are descended from one species (the wolf). The difference in behaviour in members of a genus is usually minimal, while that between different genera can differ vastly. This is one reason why it is difficult to generalise where parrot behaviour is concerned.

The second point is that with the exception of a few species such as the Budgerigar, the Cockatiel, lovebirds and Kakarikis, most parrots are not truly domesticated. Only the Budgerigar has been bred in captivity in significant numbers for more than a century. Many of the larger captive-bred parrots are only one or two generations away from their wild-caught ancestors and, in fact, there are countless thousands of wild-caught parrots in homes and aviaries. Although the importation into Europe of wild-caught parrots is now greatly diminished, especially in certain species coming from Africa, illegal importation still occurs. We are therefore basically dealing with wild animals not, as with dogs and cats, animals which have been domesticated for thousands of years.

Some species of parrots have never yet been kept in captivity. Others are quite well known as aviary birds but only a comparatively small percentage of parrot species are commonly kept as pets or companion birds. I would define these as follows:

Cockatoos	
(including Cockatiels)	8 species
Lories	5 species
True parrots	
(Grey, Amazons, etc)	15 species
Parrakeets	4 species
Lovebirds	3 species
Macaws	7 species
Conures	7 species
Parrotlets	2 species
Total:	*51 species*

Thirty of these species are from the neotropics. There is no doubt that the intelligence and adaptability of such species as macaws, conures, caiques, Amazon and *Pionus* parrots give them a special appeal to humans. In addition, they are very affectionate birds, many of whom readily transfer their affection to a human companion. This is especially true of hand-reared birds.

On the other hand, most Australian parrakeets, for example, although beautiful, do not have outgoing personalities or responsiveness to human attention, capacity for affection, which make them popular as companions. Hand-reared birds generally become aggressive to people as soon as they mature. The exception to this rule are the members of the genus *Polytelis*. It contains three elegant long-tailed parrakeets. The Princess of Wales is known for its sweet nature, thus is one of the few parrakeets which can be kept in a mixed aviary, with smaller parrakeets such as Splendid Parrakeets. The second member of the genus, Barraband's Parrakeet (called Superb Parrot in Australia) is of interest to companion parrot owners, although little known to them. Males are so gentle and sweet-natured and can become excellent mimics. These fast-flying birds, like Cockatiels, need a lot of wing exercise.

FLIGHT

The study of wing, tail and body shapes in parrots reveals a lot about their lifestyles in the wild. Few parrots have longer, more pointed wings than the Cockatiel. It is a very fast flyer and moves long distances in search of food. The long tail and slender body contribute to its speed in flight. On a number of occasions I have seen escaped cage or aviary birds flying high overhead, instantly recognisable by their calls. If a healthy Cockatiel escapes there is little or no chance of getting it back because it will travel such a long distance. I have observed many species of parrots in their natural habitat—but few have given me more pleasure than Cockatiels. I find the beauty of their flight spellbinding.

Another Australian species, the Swift Parrakeet, is named for the power of its flight. It too has long, pointed wings. Migratory, it moves between Tasmania and the south-eastern part of Australia. It has been calculated that these birds (when flying alongside a straight road over a distance of five kilometres) were flying at least 80 kilometres (nearly 50 miles) per hour. Also found in south-eastern Australia is the Little Lorikeet, a tiny 40-gram bird, which often feeds high in the forest canopy. The speed of its flight defies the imagination. A stream-lined long-tailed parrot from South America is the Patagonian Conure, which has the unusual habit of nesting in holes in river banks. It flies towards the bank at speed, closes its wings as it approaches the nest entrance and shoots straight inside: a powerful performance.

In complete contrast is the almost lazy, leisurely flight of the big black cockatoos, such as the Yellow-tailed Black. Their wings are rounded and these cockatoos almost seem to float along. But even some heavy-bodied, short-tailed species such as Grey Parrots are capable of fast flight because they possess pointed wings. In normal flight they have been observed making about five wing strokes per second and have been timed at between 63 and 72 kilometres (39 and 45 miles) per hour over a 300 metre (980-foot) stretch.

Parrots, like most birds, have evolved with an ability to launch themselves from stationary to very fast flight within the space of a second. This is often necessary to evade predators. This is why in the confined space of an aviary or room they sometimes injure themselves when they are alarmed. It is instinctive to take off quickly. This is also why some tame parrots, when unwisely taken out of doors, take off so suddenly when frightened. They can fly so far that in an instant they have lost their bearing and are hopelessly lost. It was not their intention to fly away from their owner but an act so instinctive it could not be resisted. Full-winged parrots should never be taken out of doors without the use of a harness (*see* Harness).

MOVEMENT

There is great variation in the way in which different species of parrots move. Movements are purposeful and sedate in true parrots such as Greys and Amazons. The beak is often used to give extra balance as they negotiate tree branches. In contrast, in lories and lorikeets, movements are jerky and rapid and progress is much too fast to use the bill. In some of the small lorikeets, such as Musschenbroek's, they are quite extraordinarily mouse-like when the lorikeet is running up or down wire mesh, for example. Whereas other parrots climb on welded mesh, these lorikeets scurry like a small rodent. The different styles of walking seen in parrots are most obvious when they are on the ground. The larger parrots and macaws have a distinct waddle and a rolling gait, while some lorikeets bound along in hops.

VOCALISATIONS

The various groups of parrots differ in their vocal capabilities, not so much in their talent for mimicry but in the volume of their calls. In many circumstances it is out of the question to keep very noisy parrots in the home. Whether or not a certain bird has the potential to be noisy is fairly easy to assess according to its species. Macaws, cockatoos, Amazons and some conures are the loudest. Cockatiels have very piercing calls. Small *Poicephalus* parrots such as Senegals and Red-bellieds also have quite piercing calls but they are tolerable because they are not repeated for periods of some minutes, as in the case of Amazons and cockatoos. Unfortunately, the quietest parrots are those which are not kept as companion birds: *Neophema* parrakeets such as Splendids, Kakarikis and lovebirds. Among the medium-sized parrots, *Pionus* (such as the Blue-headed), Greys and Jardine's Parrots are comparatively quiet. It should be noted, however, that Greys might mimic other, louder, birds, if they are kept in the vicinity. And any of these species might become noisy if neglected or not disciplined.

Young parrots, especially if they have been hand-reared, are not so noisy as adults. Hand-reared parrots of some species which never hear the vocalisations of their own kind might never learn them, whereas other species do not need to learn the calls which are unique to their own species; they are making them from the time they are independent.

Species with more complex vocalisations take months to perfect them. For example, Stella's Lorikeet has a peculiar contact call which is impossible to describe in words. It takes a young bird some months of practising, until it is about nine months old, to sound more or less like an adult. This is mere speculation—but could this be the reason why Greys seldom become proficient talkers before a year old, yet macaws can be talking at three months? The natural vocabulary of macaws seems less complicated than that of Greys. At nine months a young Stella's looks exactly like an adult, although it has not yet acquired full tail length. It might be very useful in a fairly aggressive species like Stella's for other members of the species in the vicinity (unlike most lorikeets, this is not a flock species) to be able to recognise a young bird by its vocalisations. In contrast, young of the large macaws are clearly recognisable by their behaviour as such even at one year old. By then their eye colour is not so bright as an adult's but their plumage is the same.

CONTRASTING BEHAVIOUR PATTERNS

For those who enjoy the study of bird behaviour, parrots are endlessly fascinating. There is such diversity in their habits, displays and preferences. There is a marked contrast in that of the parrots or groups which are very popular with companion bird owners. These are Grey Parrots, Amazons and macaws, and cockatoos.

Greys originate from Africa. They are quietly observant birds—and not at all excitable. They take stock of what is going on around them and will finally, after hours, days or weeks, act on the information gained. They have a calm demeanour and avoid aggressive encounters.

Most neotropical parrots, such as Amazons, conures and macaws, are totally different. Many species are extremely excitable; quick to react to visual stimuli; noisy, extrovert and often aggressive in their interactions with people and with other parrots. Despite these broadly similar characteristics, the various groups have their own unique behaviour patterns. Caiques (genus *Pionites*), those little white-breasted parrots which are gaining in popularity as pets, have a strange habit which I have not seen in any other species. They love to rub themselves, cat-like, against different objects. I had one which would sharpen a twig to a point, then rub himself against the sharp point. I know of a pair of caiques who rub themselves against a towel after they have had a bath! Another one enjoys rubbing himself against the leather strips on one of his toys. One Caique owner mistakenly believed this behaviour to be caused by an irritation—perhaps from a mite infestation. It can certainly be puzzling to the uninitiated.

Caiques also have a very distinctive call with which they advertise their territory. It is sometimes called 'crowing', during which they lift up the wings away from the body. Caiques are forever busy. In contrast Eclectus Parrots, for example, are much more sedate. When displaying curiosity, they will extend the neck so that the head appears to be stretched. *Tanygnathus* Parrots, such as the Great-bill, share this trait.

The body language of the white cockatoos is very different. Cockatoos come from a different region—Australia and Indonesia. They have the ability to use their crests and facial feathers to communicate with each other. The crest is raised in alarm, in display and even in play. It is also used when one bird is trying to intimidate another (or a dangerous intruder of another species); at such times the crest is fully erected, the wings are held open and the body feathers are raised in order to increase the apparent body size. This type of behaviour is especially common in Moluccan and Umbrella Cockatoos. With their huge crests they can look awesome—and their loud vocalisations add to the impression of a creature to be feared. They are among the most excitable birds in existence and, in captivity, this exuberant behaviour sometimes spirals dangerously out of control.

Sadly, many female white cockatoos have been killed by males in frenzied attacks—usually when the male is in breeding condition and the female is not compliant. This even happens with pairs which have lived together for many years. It almost seems as though for a few seconds the male has no control over his actions but is driven by a terrible urge to kill. We should not blame the male for this behaviour. The fault is always ours for keeping cockatoos in cages or aviaries which are much too small. Close confinement brings out aggressive tendencies in many parrots—but none more so than the white cockatoos. The black cockatoos of the genus *Calyptorhynchus* have a much calmer demeanour. Their displays and behaviour in general are totally different. Aggression is rare.

DANGER SIGNALS

It is important to learn the body language of the different types of parrots. If you attempt to handle a parrot when it is not in an amiable mood you are likely to be bitten, even if the bird is normally good tempered. Danger signals from a Grey are a fixed, staring look and fluffing up the head and body feathers, in order to look bigger. An Amazon which is in attacking mode will rapidly dilate the pupils of the eyes. In the more excitable species the iris is orange, so this dilating eye is like a beacon flashing. There is no mistaking the intention! At the same time the tail feathers will be flared, the nape feathers will be raised and the wings might be held a little away from the body to show the coloured wing speculum. An excited cockatoo will erect the crest, prance up and down on the perch and spread the wings and tail. Some Amazons, such as the common Orange-winged and the endangered Vinaceous, have the ability to raise their head and nape feathers when excited or when threatened.

This behaviour is taken to extremes in the striking Hawk-headed Parrot (*Deroptyus accipitrinus*) from the Amazon region. It has the most amazing feathers which can be erected in an instant to transform itself into a terrifyingly different apparition. It is not just the fan of red and blue feathers which suddenly surrounds its head, but its whole demeanour. I have been closely associated with Hawk-headed Parrots for more than 20 years. I do not know any other parrot in which aggression is so intrinsically a part of its nature. Not surprisingly, this is not a flock species but occurs in pairs—or family groups. The behaviour of hand-reared young is in total contrast to that of hand-reared adults. They are adorable, affectionate and clingy—and irresistible. The problem is that they know no fear—and the combination of fearlessness and natural aggression can be dangerous. Some do make wonderful pets until they mature—so be warned.

Excitement which is not motivated by territoriality or aggression is quite different. Many parrots will rapidly scratch the head as though the situation has produced a sudden flow of blood to that region. For example, one day I thought my lories were bored with their usual offering of fruit and decided to do something quite different with it. One of my Rajah Lories looked into the dish containing this concoction, scratched his head and bobbed up and down. I had never seen him behave in such a way to the usual fare.

RESPONSE TO TRAINING AND RE-TRAINING

While response to this varies in individual birds and according to the circumstances, the nature of the species concerned also plays a part. American behaviourist Phoebe Greene Linden comments on the desire to please and delight in praise of Blue and Yellow Macaws. She has found that even mistreated birds are remarkably adaptable to behaviour modification programmes.

They could be described as forgiving. Even birds which have

been kept in appalling circumstances have been retrained to become loving pets who trusted people. On the other hand, a Scarlet Macaw which was unfortunate enough to be subjected to any kind of abuse, verbal or physical, would react in a very different manner, meeting violence with violence. It would be much less likely ever to trust people again. The fact that abuse occurs, either out of ignorance of the emotional needs of a parrot, or intentional abuse, cannot be ignored. There are kind-hearted people who want to take on such birds and enquire whether such parrots can ever become loving companions. The characteristics of the species would be equally as important as the bird's past history in influencing the outcome. Some species are naturally more aggressive and assertive than others.

Part Three

WHY DOES MY PARROT . . . ?

In the following pages problems which are frequently encountered by parrot owners are examined. The advice offered aims to help them understand why the problem has occurred and how it can be rectified. It is emphasised, throughout the text, that a multitude of circumstances cause any one form of behaviour and the potential for correction.

WHY DOES MY PARROT...

Attack me?	81-85
Avoid the sun?	174
Bite me?	89-92
Behave aggressively?	76-80
Crouch down and bob his head?	110
Dip his head in the water dish?	89
Dislike having his head scratched?	146
Dominate me?	105-106
Dunk his toast?	119
Fall off his perch?	133, 143
Feed me?	156
Fly at people and bite them?	81-82
Gnaw the furniture?	104
Go through a 'difficult' period?	75
Grind his beak?	89
Growl?	111-112
Hate his cage?	97-98
Hate me?	75
Imitate the telephone?	135
Lay eggs?	131-133
Lie on his back?	87
Nibble my hair?	147
Not talk?	176
Pluck his feathers?	112-117
Prefer my husband?	95, 166
Raise a foot when I go near?	111
Refuse to 'step up' when on my shoulder?	167-168
Scratch on the floor of his cage?	98-99
Scream?	154-162
Sleep with one eye open?	168-169
Tear up the paper on the cage floor?	164-165
Understand human speech?	179-180

A

Adolescence

The adolescent stage in hand-reared parrots kept as pets is such a contrast to that of the recently weaned parrot that it comes as a shock to many parrot owners. They are simply not prepared for it and often do not recognise it as adolescence. 'Why does my parrot suddenly hate me?' is a common reaction of the owner. He doesn't 'hate'. He is starting to challenge your leadership—just the same behaviour as one finds in most human adolescents. Before he was affectionate and compliant; now suddenly, his behaviour changes.

In parrots this occurs at different ages. In small species it could happen as early as six months and in Senegals at about nine months. In long-lived birds such as large macaws, it will usually occur at two to three years. A change of behaviour will be noticed as hormone surges commence.

A species which often goes through a very difficult period is the Senegal Parrot. It can become extremely nervous and absolutely refuse to be handled. It is important not to lose one's patience or to lose interest in the bird. Ask all members of the family to talk to him and praise him when he is behaving well. Look on this as a trying period, after which he will hopefully emerge as much more sweet tempered.

Many parrots are difficult to live with during the adolescent stage because they start to challenge their owner's authority or dominance. Parrot behaviourist Liz Wilson was boarding a nine-month-old Grey Parrot called Freddie. Sometimes when she put her hand into his cage and said: 'Up!', he would put his head down as though asking her to scratch it. Alert to his every move, however, she noticed that he was looking at her out of the corner of his eye. A truly submissive Grey, soliciting head-scratching, would not do this: it would close its eyes or look down. If she fell for the head-scratch, he would bite her. Why? He was not malicious. He was challenging her authority. By outwitting her into allowing him not to carry out her command, the Grey had established that he was dominant over her.

How did she overcome this attempted manipulation and remain on good terms with Freddie? She out-smarted him in this way: as she said 'Up, Freddie!' and pushed her right hand against his lower abdomen, she waved her left hand in the air. This sudden distraction would cause him to forget what he had been about to do and to step on to her hand. Greys being clever birds probably would not fall for this trick too often, but it is a good one to keep in mind!

The onset, duration and difficulty of adolescence vary in individual birds. In some the period is hardly noticeable. In others it is a testing time. While the owner's attitude and experience will influence behaviour during this time, the parrot's individuality will also be significant. Training and frequent reinforcement of commands, such as 'Step up!', will help the owner to re-establish his or her authority.

Aggression

The basic reason for aggression must be established. The cause could be due to one of the following:

1. Like dogs, some parrots will use aggression to establish control over their owner. Biting is an efficient way of doing this. Hand-reared birds, being fearless, often resort to these tactics.
2. Aggression, that is lunging and biting, may be the only means of defence for a truly frightened bird in a confined space, such as a cage.
3. Aggression often occurs because a parrot is defending what it perceives as an area which is its own territory. Or it is defending its favourite human within that territory.
4. A behaviour which appears to be aggression is seen in some young hand-reared parrots which are hungry. Forced weaning is to blame. An attempt to wean a parrot too early can

leave it anxious and permanently hungry because it is not yet able to eat enough (or enough of a suitable food) on its own.

5. Aggression can be inherited. Some breeders choose their breeding birds for size or colour, but those who are breeding species for the pet trade should be selecting birds which have a good temperament. A classic example of inherited aggression among birds in my care concerned Cuban Amazons (*Amazona leucocephala*). The male of one pair was so aggressive that he had to be removed from the aviary just before the young fledged. In the first year he killed his sons soon after fledging. Ultimately, after many years of breeding success, he also killed his female. I found her and I saw the head injuries inflicted. This was a sad end to a lovely bird. Several years later one of his sons was old enough to breed. After a couple of seasons he killed the female. Again, I had the misfortune to pick her up. It chilled me to see that the head injuries which had caused her death exactly resembled those which his father had inflicted on his mother. In my opinion, parrots from a known aggressive line should not be used for breeding. But because aggression is not a tangible quality, like size, it is ignored by breeders.

Some parrot species are naturally more aggressive than others, such as cockatoos, certain lories and certain Amazon parrots. Some species are naturally timid and seldom resort to aggression among their own kind. However, aggression in pet parrots is a problem that has increased with the number of hand-reared birds. This is because such birds grow up without a fear of humans. In such birds, aggression can be a form of exerting dominance, usually in birds which have never received any training or discipline. Among wild-caught and parent-reared parrots aggression is usually a manner of expressing fear. If, for example, a wild-caught parrot is kept in a small cage, it will strike out at an approaching hand because it cannot move away from this threatening object. In an aviary, where the bird could fly off or move away, it might appear timid but would not behave aggressively. Thus, in this case, aggression is used as a form of self-defence.

Question

I purchased an African Grey Parrot six weeks ago, when she was 13 weeks old. She is a very aggressive bird and does not usually 'listen to' anyone apart from my husband. She bites a lot and is trying our patience in attempting to tame her. She does not pluck her feathers. I have taken her to a vet who says she is healthy. We had her wings clipped so that we could let her out as she was crash-landing before. She has a big cage and we have built her a climber so that she can play. She even has a play pen now. Two days ago we bought her an Amazon as a companion. He is just one week older than she is. When we let her out she goes to his cage and tries to bite his eyes or legs. We dare not let her alone with him. We feed her first so that she does not get jealous—but nothing seems to help.

Answer

I must be honest for the sake of your Grey, and tell you that the problems you are experiencing cannot be blamed on your parrot. A young Grey is not naturally aggressive. Fear and aggression are born of mistrust, mishandling and even hunger. First of all, a concerned seller, breeder or dealer would have made sure that you were well informed and that you had a book on the subject. The seller should also have been able to offer you advice which would have prevented the problems you are now encountering.

I note that your Grey was 13 weeks old when you obtained her and was fed mainly on sunflower seed. First of all, when a grey has been moved to a new home at 13 weeks, it is advisable to spoon-feed it twice a day for at least the first few days to ensure that it has sufficient food in its new surroundings. Greys are nervous and may not feed well at first even when adult, let alone when they are not properly weaned. At 13 weeks no Grey is eating enough of a hard food like sunflower seed. It is better to offer it soaked to such a young bird. (Provide it twice a day in warm weather to prevent mould forming on it.) It needs a lot of soft foods—fruits, vegetables, wholemeal bread, pellets soaked in fruit juice, etc. My guess would be that your Grey was hungry initially, and this made her anxious and even aggressive. She was force-weaned. This can affect a parrot's temperament for months to come.

The next mistake you made was to have her wings clipped. This would make her even more anxious and fearful. If she was crash-landing before, it was because she was very young. It takes a few weeks for some Greys to perfect their landing skills. Also, they need perches in the room as obvious landing places. If a young parrot's wings must be clipped (*see* Wing-clipping), it should be done gradually—not overnight. Surely she is more likely to fall and crash-land with clipped wings? A young bird cannot develop its pectoral muscles if its wings are clipped at an early age.

The first step to correct the aggressive behaviour is to teach the basic commands, such as to step on to your hand. Aggression is often the result of the parrot believing he or she is dominant over you—because it has never been disciplined properly. If this problem is not rectified before the age of six months, it will be difficult to deal with in the future. If you or your husband are now afraid of your Grey, because she bites, taming her will be impossible and it would be better to let her go to someone who understands how to win her trust. It worries me that you write: 'She bites a lot and is trying our patience...' Taming a bird, winning its trust, can be carried out only with patience and love and can take a long time. There is no quick method. Only when she has confidence in you will her aggression subside.

I am sorry that you did not seek advice before buying another bird. This is the worst step you could have taken. She needs all your attention. She needs reassuring that you love her. Now you are giving your attention to a newcomer. I am not surprised that she tried to bite the Amazon. It is unfair on both birds. The Grey will be extremely jealous and this could make her even more aggressive. The Amazon does not need the stress of an aggressive companion in a new home. The Grey does not need a parrot companion. She needs a lot of attention and very skilful and sensitive treatment. Greys are, in fact, too complex emotionally to make good pets for inexperienced parrot keepers.

My advice would be to take the Amazon back—but that is probably impossible now. All I can suggest is that you keep them in separate rooms when you and your husband are in the house and spend a great deal of time talking to them, offering them titbits of food and winning their confidence without trying to force the taming process.

Aggression in Cockatoos

One parrot behaviourist commented on the large number of calls she received enquiring if someone would take their aggressive or unmanageable cockatoo. Could it be put into a breeding situation, put on drug therapy, neutered (!) or even euthanised? She believed that the aggression described was often the need to achieve and maintain a certain position within the 'flock', regardless of whether the flock consists of birds or humans.

In the latter case, the person who handles the bird may unknowingly be initiating that aggression. If the cockatoo is out of control—because it has not been disciplined—the reaction of some people is to try to dominate the bird with loud or threatening actions. The cockatoo is challenging this reaction with aggressive behaviour. On the other hand, aggression may stem from a strong feeling of territory or the need to protect his favourite human from other humans. It could well result from jealousy, especially when other people are diverting the attention of the favourite person from the cockatoo.

If aggression is not the result of fear, the best advice is not to handle the cockatoo when it is in a dominant or territorial mood. It is thus necessary to learn to read the cockatoo's behaviour. By carefully observing it—but in a casual manner, avoiding direct eye contact—a pattern of behaviour will emerge which acts as a warning of impending aggression. This aggression might be averted by ensuring that your own behaviour is low-key, non-aggressive and nonthreatening. Maintain a relaxed body posture and facial expression. Never be loud; talk quietly and confidently to yourself, without looking at the cockatoo. Never let him think that he has won a confrontation by chasing you away. If necessary, leave the room before he becomes aggressive. Never handle him when he is in an aggressive mood. Learn which time of day this is most likely to occur.

Anthropomorphism

The dictionary definition is 'attributing human form or personality to animals'. Many parrot owners attribute a way of thinking which is more human than avian to their birds, and this may cause them to misinterpret their actions. On the other hand, it is important to understand that parrots' emotions can be compared with ours, in as much as they can experience a strong emotional attachment (love) to another parrot or to a human, or the reverse, dislike, and there is no doubt that they can know jealousy. By their actions they can indicate that they show concern for another member of their species. To recognise that they can feel these emotions is not to anthropomorphise but to understand what is true.

I do find it irritating when parrot owners refer to themselves as 'Mummy' or 'Daddy' in relation to their parrot—and the parrot is 'my baby'. A parrot is still a wild animal and it has dignity. To treat it as though it is a feathered doll is to strip it of that dignity. Especially for someone who lives alone, a parrot can be an extremely important companion—but that is no reason to treat it like a child substitute. Unless it can be respected for the remarkable species it is, inevitably it will be misunderstood.

Attack

Attack is the most dramatic and explosive form of aggression. It is not uncommon in more volatile species, such as cockatoos, Amazon parrots and even conures. Because it is a real threat to the safety of the person being attacked and many parrot owners do not know how to deal with it, it is often the reason for parrots being sold, given away or sent to rescue centres. An attempt must be made to resolve the problem after the first attack—because it will only get worse. (*See* also Aggression.)

Question
My two-year-old male Umbrella Cockatoo spends most of the day out of his cage. He has the freedom of the house. Unfortunately, he has started flying at people—my boyfriend, my father and my brother, for example—and biting them on the face. He has never done this to me. What can I do to stop this?

Answer
As your cockatoo matures, this behaviour will get worse unless you take steps to stop it now. You cannot expect your family to

tolerate this. A mature male cockatoo which has not been trained or disciplined is, frankly, a dangerous animal which could inflict a serious injury on someone. It is not the cockatoo's fault. The bird must be disciplined.

In all probability your cockatoo considers itself bonded to you and, in attacking people, he is warding off competition for your attention. After two years at liberty within the house your Umbrella Cockatoo is not going to take kindly to being confined to his cage, and this may result in screaming and seed-flinging for attention. I would suggest that you buy or construct a large indoor flight cage and keep it filled with fresh-cut branches, plus a selection of toys. Perhaps you could take him to your bedroom for 'time out' for set periods every day, so that he does not come into direct contact with other household members. Also, away from his cage or his flight, his territoriality may diminish. But you must lavish attention and affection on him on a daily basis or he will become loud and insecure and may start to pluck himself.

Confining him to an indoor flight might not work because screaming for attention might be a problem. However, I am not sure that wing-clipping is the answer. Although he could then spend more time out of his cage in the company of family members, he might still try to attack them by biting their ankles. But this is less dangerous and more readily dealt with (by wearing boots, perhaps), than facial attacks.

As well as implementing these changes, it is imperative that you start to train him. Once he responds to various commands, his aggression should lessen, as he accepts that you are in control of the situation.

I believe that it is a fundamental mistake to let a parrot have the freedom of the house for most of the day. It will react badly to being returned to the cage. The reverse should be true. The cage is its base and time out is special. Is it any wonder that it becomes territorial and aggressive when other people enter the house? It has come to regard the house as its personal space and other people as intruders.

Question

Four months ago my husband bought me a Lesser Sulphur-crested Cockatoo aged three years. The cockatoo's manners

were good and he settled in really well. He has the freedom of the house while someone is at home. I feed him, talk to him and tickle his head and deal with all his needs. He has fresh fruits and vegetables, carrot and sweetcorn being his favourites. During the past two weeks his behaviour has changed. It is obvious that he has become my husband's bird and he shares my husband's dinner. He also enjoys the company of my elderly father on his twice weekly visits and he stays on my father's knee or shoulder all day.

Last week as I walked past my father Cocky flew at me and gave me a nasty bite on the back of my neck. When my husband arrived home Cocky flew at me again and bit me on the side of the head so hard that I almost went over. In the past I had him on my shoulder for hours as I did the housework or prepared food. He even learned some words in my voice. Although my husband and I have been married 36 years we often sit together and often kiss and cuddle. Could Cocky be jealous? I am really fond of him and do not want to sell him or to keep him caged. What can I do?

Answer
Cockatoos form very strong pair bonds—stronger than almost any other birds. Yours has, in effect, bonded to your husband. You are an intruder into his territory and a competitor for your husband's affections, therefore he attacks you. At three years old he has become sexually mature and this has aggravated the problem. Mate protection is such a basic element of behaviour it will be difficult to alter. As in the case described above, training offers the best hope because training instils respect and diminishes aggression. However, whether the cockatoo would then allow himself to be handled by you is difficult to predict. While he might respond to your 'Step up!' instructions in your husband's absence, he might not tolerate you in your husband's presence. It is difficult to predict what will happen, as a lot depends on the reactions of the people concerned. You seem very forgiving and, despite the bites received, not afraid of the cockatoo. In your place, many people would be afraid and the cockatoo would understand this from your body language. This knowledge would probably make attacks even more likely.

   The main problem, as I see it, is that it has become unsafe to allow the cockatoo its freedom during the day and confining it would lead to further problems. The advice given in the case above applies equally here. You might be considering building an outdoor aviary, and obtaining a female Lesser Sulphur-crested Cockatoo. However, this course of action is fraught with danger, as some male cockatoos, more especially those which have been hand-reared and never socialised with their own kind, attack and kill females. This strategy is much more likely to be successful in parent-reared (or wild-caught) males. It might be borne in mind, however, that the Lesser Sulphur-crested Cockatoo is now an endangered species, due to over-trapping for the pet trade. Captive breeding is only of value

where the parents are allowed to rear their own young to independence and these young are used for future breeding
purposes. If they enter the pet trade and they are males, they
might ultimately kill more than one female—because the bird's
history is not passed on when it changes hands.

B

Bach remedies

During the 1920s an English doctor, Edward Bach, developed the idea that, in humans, illness can be used to restore harmony between body and spirit. Using the essence from 38 different flowers and trees, he demonstrated how negative emotional and character conditions can be reversed. He used such trees as willow, elm, larch and walnut and shrubs such as olive, wild rose, crab apple and clematis. The plant extracts are diluted with spring water and conserved in alcohol. The essence concentrates which result can be obtained at homoeopathic pharmacies, from health food stores and from specialist mail order companies. Their great advantages are that they produce no harmful side effects and are not expensive.

There are several suggested methods of administering them to birds. One is to add one or two drops of concentrate to the drinking water. However, most parrots drink so little that this might not be effective. The other method is to dilute one or two drops of the Bach flower with one millilitre of bottled spring water and drop it on to the bird's head or under the wing. Some homoeopathic remedies are available by injection or as ampules (in a saline solution).

It is known that negative psychological conditions cause disease and prevent recovery in humans, animals and birds. Bach flower therapy is effective for treating some behavioural disorders and psychological problems in birds. Some feather pluckers and anxious and aggressive birds can benefit greatly from this treatment. As yet, most vets have little knowledge of homoeopathic remedies. It is necessary to be referred by a vet to a homoeopathic vet. Some use a combination of traditional and homoeopathic methods.

Case history

A Citron-crested Cockatoo plucked and mutilated (made a hole in) its breast. Valerian was given for nervousness (the dose was worked out by the vet) and Vitamin E was given for healing. After several days the bird stopped plucking and mutilating itself. After five months it was in perfect condition again.

Back, lying on

Some parrots seem positively to enjoy being on their backs. This can be very alarming to a person who is unaware of this characteristic and who sees a parrot on its back, feet in the air, on the cage or aviary floor. Young hand-reared macaws adore being turned on their back and cuddled and played with. Most Grey Parrots would not tolerate such an indignity after the age of weaning. Lories will also play on their backs; when they are young some will lie on their back and nibble their toes. Chicks, even before they are feathered, often sleep on their backs.

Lying on their back is also a defensive behaviour in some species—mostly neotropical parrots. Chicks of parent-reared macaws instinctively turn on their backs when threatened, and lash out with the feet. Even chicks have powerful legs and sharp nails. When cornered, some Amazons and lories will also turn on their back. Their feet, with their powerful grip, are weapons that are nearly as formidable as their beak.

Bathing

Most parrots are enthusiastic bathers. The exceptions are some species which originate from very arid areas. These are in the minority; there are far more from rainforests. In some regions there are very heavy but not prolonged rain showers every day. The atmosphere is humid and this humidity assists in plumage care. To me, there are few sadder sights than a parrot which has never been showered or permitted to bathe. The plumage looks dry and dusty—and the parrot's condition spells neglect.

It is best not to give your parrot access to a bath in the

evening, to prevent him from going to roost with wet plumage. There are three main methods of ensuring that his plumage receives a good soaking, at least once or twice a week, or more if this is indicated.

1. Use a plant mister filled with warm or cold tap water; set the nozzle to a fine spray. At first (especially with young parrots) spray lightly, but when spraying becomes a regular and much enjoyed event, you can soak the plumage. Some parrots will anticipate a shower with excitement, spreading their wings in expectation of a good soaking.
2. Some owners of the larger parrots take their bird in the shower with them. This is not advocated for young parrots, which should not receive a drenching shower before the age of about six months. Go carefully at first, to avoid frightening the parrot. You might let him watch you having a shower a couple of times. The running water may induce him to call out and ruffle his feathers. This is a sure sign that he wants to get wet! Showering in this way can be quite convenient because a bathing cockatoo or macaw can splash a lot of water in the vicinity. This method however is not without its risks. Apparently one macaw which was taking a shower with his master, lost his footing. On the way down he caught hold of a conveniently placed appendage!
3. Place a large but shallow bowl of water on the floor, or on the cage floor if the door is large enough. A leaf of spinach or lettuce floating on the top may encourage your parrot to take a dip.

The benefits of regular showering will be apparent in glossy plumage and well lubricated feet and skin. Above all—and this is especially important with cockatoos and Grey Parrots—it prevents the build-up of powder down. This is the white dust which a parrot gives off when it shakes its plumage or flaps its wings. This dust can have serious consequences for asthma sufferers. Other birds in the vicinity might also be allergic to it.

Question
Why does my Blue-fronted Amazon Parrot often dip his head in his water dish?

Answer
He wants a bath. The water containers in most parrot cages hold enough water for drinking only. Your Amazon is so desperate to feel water on his plumage that he is doing this the only way he can. Amazons are among the most exuberant of all bathers. Don't deprive him of this important activity.

Beak clicking

Rapid beak clicking is a form of behaviour seen in a variety of species. It is often a threat behaviour or a warning that the intruder is too close. At the same time, the body language will clearly indicate that the parrot is uneasy and probably preparing to lunge or bite. However, it is not always a threat. Some parrots beak click in excitement. The lack of posturing or aggressive behaviour will be the clue to the motive.

Beak grinding

A question which I have been asked on numerous occasions over the years, is: 'Why does my parrot grind his beak?'

The grinding together of upper and lower mandibles when a parrot is quietly resting, often at night, is a natural act, carried out only when the bird is completely relaxed. It is often suggested that it occurs to keep the mandibles in good, sharp condition. My own guess would be that it was derived from the need to keep the beak free of particles of food (some are fibrous or sticky), then developed into a comfort activity.

A parrot's beak grows continuously, like our fingernails. Both are made of the same material: keratin.

Biting

This is one of the most common problems with companion parrots and usually is derived from the owner's lack of understanding of his or her bird. The problem needs to be examined on two fronts: that which relates to a parrot which has been hand-reared and that of a parrot which has not been tamed. One is biting to try to establish dominance and the other is biting out of fear (*see* Training).

Hand-reared Parrots

Young parrots are very playful and, like puppies, will test objects by biting at them. These objects include fingers. At first the parrot does not bite hard; it is just curious to find out what fingers are. Then one day it bites really hard and elicits a yell from its owner. To a parrot, this seems like a positive response. It is fun. Right from the start, the new parrot owner should discourage his or her bird from playing with fingers. But if one day the bird bites anyway, be prepared for this and make an effort not to respond in any way. If there is no reaction to biting, it is not much fun. Normally gentle parrots sometimes bite when they are playing. This is usually because they become over-excited—not because they are malicious.

At other times a parrot might appear to bite when, in effect, mishandling has caused it to lose its balance. One parrotlet owner described how the parrotlet would steady himself with his beak as he stepped on to any perch, including a finger. Someone new who was unfamiliar with this action might interpret it as a prelude to biting and might pull away. If the little parrotlet already had his beak in position to step up, he would hang on to the finger to regain his balance. The person holding him would then believe that he had been bitten. The moral here is surely either to explain this to the new handler, asking him to keep his finger firmly in position, or not to allow strangers to handle such a small bird.

As a parrot matures, he will start to challenge your authority. Refusal to comply with commands, and biting, are symptomatic (*see* Adolescence). Training your parrot to respond to basic commands, such as 'Step up!', gives you a much greater degree of control. As your parrot accepts that he should respond to the commands, his respect for you as flock leader will grow. A parrot does not bite a flock member of senior rank.

This is all very well in theory, you may say, but dealing with a biting parrot is not easy. It might give you more confidence to use very thin cotton gloves, such as the cotton anti-allergic gloves sold in some chemists or drugstores. The gloves should be of a neutral colour. Better still, train a biter to step on to a small perch or stick, rather than on to your hand. If he wants to bite at the stick, this does not matter. It dissipates some of his aggression.

All too often wing-clipping is advocated as a means of dealing with a full-winged bird which is biting. The basic problem, however, is not that the bird is full-winged, but that it has not been trained and disciplined. Clipping its wings when it has been used to freedom, will not improve its quality of life. Training it to respond to the most basic commands will.

One also has to accept that parrots, like people, have their idiosyncratic ways. My Amazon would bite me if I attempted to pick her up when I was wearing an outdoor coat or jacket. It was difficult to explain why. I just had to accept this and avoid such confrontations.

Any change which destroys a parrot's confidence, whether it is actual abuse or even wing-clipping, can cause a parrot's personality to change to the degree that it starts to bite. A reader of *Parrots* magazine described what happened when a Grey was wing-clipped at the age of six months. After being clipped, he could fly in a straight line but could not gain height. Subsequently, he broke a couple of tail feathers and was unable to fly at all. Then his character changed from affectionate and playful to being a moody biter. Previously when he was startled he would have flown away—but now he bit, instead. Yes, parrots do bite if they feel threatened and are unable to escape from danger, perceived or real. When this Grey's feathers grew back, he slowly regained his confidence and stopped biting. His owner said that he would never again clip a parrot's wings.

Question
My 15-month-old hand-reared Green-cheeked Conure has started to bite and behaves like a timid wild bird when I go near the cage. Why is it doing this?

Answer
It is not unusual for the temperament of pet conures and other parrots to change at adolescence, especially during the breeding season when hormones may influence behaviour. Many hand-reared birds go through a phase of nipping when they reach adolescence. However, it is not usual for a bird to behave as though it is afraid; normally the behaviour at this stage is just

the reverse—trying to become dominant. I suspect that your bird has nipped someone who has retaliated, perhaps frightening or hurting it. A bird which is treated in this way may take months to regain its confidence—if it ever does. I think you should try to investigate the attitude of other family members towards your conure. Any form of abuse must be stopped. *See* Phobic Behaviour.

Blindness and cataracts

One of the factors that can radically alter the behaviour of a parrot is blindness or failing vision. Many Budgerigars suffer from tumours as they get older; tumours in the pituitary gland cause pressure on the nerves leading from the eyes to the brain. This can cause blindness. As in humans, cataracts occur when the cells in the lens of the eye degenerate and die, causing the lens to become opaque. In rare cases, cataracts have been removed from the eyes of the larger parrots.

Question
Why does my Budgie now spend most of her time on the cage floor? She eats well, although she is nearly 12 years old.

Answer
I would suspect that she has gone blind or that cataracts are obscuring her vision. If you move your hand up and down in front of her face and she does not move, it is unlikely that she can see your hand. Also move the food container and watch what happens. Blind birds find the way around the cage out of habit. If she goes to the place where it used to be, you know she is blind.

Blushing

Question
I have a Blue and Yellow Macaw and notice that sometimes the bare area on his cheeks becomes very pink. Why is this?

Answer
The bare area on the cheeks of macaws and a few other parrots are covered in tiny blood vessels. When the bird becomes

excited or angry, the cheeks "blush" red—but not for the same reason as an embarrassed human! In macaws, the reason might be pleasure at being greeted by a favourite human. This does not happen in Grey Parrots, in which the bare area is confined to the lores and the area around the eyes.

Boarding

When you go away, who will look after your parrot? This is a question which every potential parrot owner should ask themselves. It would be wrong to think that you can leave your parrot alone in the house and ask a neighbour to pop in and feed him. Without human contact and company, a parrot might refuse to eat or pluck its feathers or mutilate itself. The stress could have a long-term impact. If your house will be empty, the best solution would be to board him with someone he knows and likes. This is impossible for many people so they have to pay for care in another person's home.

Unlike boarding establishments for cats and dogs, anyone can board parrots, simply by advertising the fact that he or she will do so. You might be desperate to find a temporary home for your bird and start by enquiring at your local pet store. Some pet shops do actually board birds. However, this is not an ideal situation, especially if they buy and sell birds. The disease risk to your parrot is enormous. The same problem applies, probably to a lesser degree, if you contact someone who boards parrots in their own home. You might locate such people by looking in the advertisements in *Parrots Magazine* (a monthly publication) or by contacting the local parrot society or cage bird society. However, you should visit the person who operates this service before agreeing to leave your much loved companion in his or her care. Don't expect to find someone on your doorstep. You might have to travel some distance.

If the person seems suitable, take your bird in its normal cage (it needs something familiar in its life), with full instructions on feeding and habits. Some parrots settle down very quickly in new surroundings, especially if they like the person caring for them. Others will be withdrawn and suspicious and may not feed well for a few days. Leaving a supply of favourite tit-bits is therefore useful.

Bonding

A parrot is said to be bonded to a person or to another parrot when it has a very close relationship with him or her. It is a myth that in order for a person to achieve this it is necessary to hand-rear the parrot which is to be the companion. Bonding can occur at any age if the circumstances are right. It basically means that the parrot trusts you and prefers your companionship to that of other people or birds.

We should not assume, however, that all parrots want to bond to a person. Some don't. They would rather be with another parrot. If we try to force our attentions on such birds, it will have no effect, except to make the parrot wary. Sometimes we must accept that a parrot will never be successful in a pet situation. If it is suspected that it would prefer to be with its own species, its reaction to the presence of another parrot should be tested. This does not necessarily mean placing the parrot in a breeding aviary. It might be perfectly happy in a home, as long as it can live with another parrot.

There is a commonly held belief among parrot owners that a female parrot will always prefer humans of the male sex and that a male parrot will prefer the ladies. This is true of many parrots—but it is by no means always the case. Regardless of sex, one problem is that a parrot forms such a strong attachment to one person that it merely tolerates other family members. The aim should be to prevent such a situation arising by allowing as many people as possible to handle the bird—right from the beginning (*see* Training).

Despite a strong bond with one human, however, young adult parrots can change their allegiance. In the USA Jane Hallander carried out an internet survey among owners of Grey Parrots. The number of birds involved was not given, only the fact that there were fewer than 100. The survey revealed that 63 percent of male 'Congo' Greys (the name is sometimes used to denote birds of the nominate race) changed their preference from the primary care-giver to another family member. This happened when they were aged between two and three years. This is interesting as this is probably the age at which they would start to look for a mate in the wild. Until then, a sibling could have been a constant companion. In female 'Congo' Greys, 16 percent of the birds in the survey changed their human allegiance. A surprising fact was revealed by this limited survey. No Timneh Greys (*Psittacus erithacus timneh*) changed their 'pair bond' (Hallander, 1999).

Question

Our five-year-old female Red-bellied Parrot (*Poicephalus rufiventris*) was given to us when she was one-year-old, as she was too attached to people to be used for breeding. From the outset, she bonded so strongly with my husband that it is impossible for other members of the family to handle her. She can be aggressive. What can I do to win her affection?

Answer

Nothing, in all probability. It is too late. The bond between her and your husband is too strong now. She regards other family members as competition for your husband's attention.

As long as he is around on a permanent basis, her behaviour is unlikely to change. My own Amazon parrot, my companion of 15 years at the time, could never be handled by anyone else. However, for a few months, I was working away from home during the week, returning only at weekends. During this time she became much friendlier towards my partner, because he was the only person in the house. Indeed, I still have a photo which testifies to this, of her sitting on his lap and eating from his plate. This is the only time during the 39½ years she was with me that she could be handled by anyone else. This proves that under certain circumstances a parrot's behaviour can change.

Indeed, it can change in a way which is not desirable—as was discovered by a lady who had a Blue-fronted Amazon. During her absence of four months, her Amazon stayed with a friend. When the move to her new house was completed, her Amazon returned to live with her. However, its attitude had changed totally. The bird was aggressive and unfriendly. This was heartbreaking as the Amazon was her sole companion. She sought advice. What should she do? In such a situation the answer is to have patience and to try to act as though nothing is amiss. Above all, try to avoid confrontations which might lead to the parrot biting. She followed this advice and she persevered. It took several months for their friendship to be totally re-established but her sympathetic approach resulted in success.

Sometimes there is no alternative but to find a temporary home for a companion parrot; however, the period should be kept to a minimum, where possible. The longer the bird has to become attached to a new person, the more difficult it can be to break that bond.

Breeding condition
See Sexual behaviour.

Buttons

Question
Why does my macaw always try to chew the buttons off my shirt? And how can I stop him?

Answer
To a macaw, or any other parrot, buttons are just miniature toys, waiting to be picked off like little fruits from a tree. Furthermore, they make a satisfying cracking sound when beak pressure is applied. You might try to train him to leave buttons alone – but they are such irresistible objects, I am not sure how successful this would be! When he reaches for a button put him down at once for a minute without any cross words, then come back and pick him up. Keep walking away when he reaches for or removes a button. Personally, I would never wear anything with buttons (or jewellery) around a tame parrot as I think the temptation is too great to be resisted. Avoid the problem by wearing a T-shirt.

C

'Cage-hate'

I firmly believe that every companion parrot should have a cage for its base and sleeping place. Some parrots, especially young birds, are initially allowed as much freedom as they want outside the cage, perhaps on a play stand. Naturally, they become reluctant to return to the cage. However, for reasons of safety and during periods when normal routine is disrupted (perhaps when the house is being redecorated), it is necessary that a parrot is accustomed to being confined in a cage on a daily basis. In addition, a parrot which has unlimited freedom will come to believe that it is in charge of its own life. This will lead to a multitude of problems, those that stem from lack of discipline. Imagine, too, if it takes a dislike to a certain visitor to your house, it would be free to attack him or her at any time. It might seem unlikely that anyone would want to keep a parrot free, especially in view of the high risk of escape (whether or not the parrot is wing-clipped)—yet there are people who chose to keep their bird in this way.

If a parrot is reluctant to return to his cage, make sure he receives no tit-bits when he is free in the room. Reserve tit-bits for when he is inside its cage. Placing them inside when you want him to return will not pass unnoticed. It will be a big incentive for him to go back. Be on hand to shut the cage door at the appropriate time.

Another incentive to return to the cage might be little cardboard boxes filled with food treats. One can buy tiny boxes of sultanas, and these are just the right size for a parrot to hold in its foot. Place one inside the cage before your parrot enters, and vary the treats it contains, and your parrot might even look forward to entering its cage!

Question

I had a Blue-fronted Amazon for 20 years. Sadly, he died recently so I decided to buy an English-bred hand-reared Yellow-naped Amazon. We have had him for eight weeks. He has not taken to being in his cage, always looking for a way out. Getting him to settle on top of his cage is impossible. He just wants to fly off. Will he calm down as he gets older?

Answer

Hand-reared birds are very different in behaviour from the imported bird you bought 20 years ago. They are much more demanding and constantly clamour to be let out of the cage, especially if there are no set times for this. Also, it often happens that when a bird is first obtained it is spoilt by being allowed out of its cage for hours on end. Then gradually the time is reduced and the parrot objects.

I would suggest that you take him out at set times, so that he gets to know when this will be. Birds have a very good sense of time and will soon learn the routine. Perhaps you could have him out at least twice a day. In the meantime ensure that he has toys to play with and willow twigs to chew up when he is inside. Regarding not wanting to stay on top of his cage, you cannot expect a hand-reared bird to do this. You might consider making or buying a 'play gym'—a stand with ropes and swings and toys on it. This will help to keep him amused and to give him a focus of attention when he is outside the cage.

Question

I bought a recently weaned young Grey Parrot four weeks ago. She seems to hate being in her cage. As soon as she is put back inside she goes down on the floor and starts to scratch vigorously like a chicken. Why does she do this?

Answer

This strange habit of scratching with the feet on the floor seems to be confined to Grey Parrots and members of the other

African genus, *Poicephalus*. Virtually all Grey Parrots do this when they are very young! Fortunately, they soon grow out of the habit. Make sure that there are plenty of items within the cage to keep your Grey busy, such as strips of leather tied to the bars.

Children and babies

Children must receive proper instructions for handling parrots. If they are too excitable or too fearful they should not be permitted to do so. Very young children should not be allowed to handle parrots, both for their safety and because their quick and inconsistent behaviour will frighten many parrots and make them afraid of being handled by other people. Also, can a child be trusted not to tease a parrot? Sometimes children want to put something in a parrot's food dish. Instead of approaching calmly, they make fast, snatching movements. These are likely to be met with aggressive behaviour from the parrot.

Conversely, can a parrot be trusted not to harm a child? No risks should be taken. I once saw a visitor to a bird park sitting on a seat with a very young child. A large macaw was sitting on the edge of the seat. The unbelievably foolish mother took the child's finger and put it in the macaw's beak. The macaw could have amputated the finger. Fortunately, it did nothing, but annoying even the most gentle bird in this way could have caused it to bite out of fear.

It seems that many parrots are aware of the vulnerability of human babies and will not bite them. A parrot in a family known to me could be trusted only with the father. It would have bitten the other family members, given the opportunity— with the exception of the baby.

Sadly, the arrival of a new baby in a home where there is a parrot often results in the bird being banished. This is especially true of cockatoos. The parrot becomes noisy and is described as being jealous of the new baby. In truth what happens is that the parrot or cockatoo is neglected. It is just calling for attention. Careful thought should be given to such a situation before a parrot is purchased. If there will not be time to maintain the standard of care and attention, this is not a household where a parrot will be happy.

Case history

An Umbrella Cockatoo lived with a family who had two children who were some years older than the cockatoo. When she was three and a half years old another child was born. The cockatoo became very noisy and was inhumanely banished to a shed in the garden. After two months in solitary confinement the owner contacted the breeder who agreed to buy her back. On arrival the breeder was shocked by her appearance. She had plucked all the feathers she could reach. Only her head was feathered; her body was covered in white down and was totally lacking in contour feathers. Fortunately, back in a caring household, her plumage slowly improved.

Coughing

If this occurs on isolated occasions, without repetition, there is unlikely to be cause for concern. If it is frequent in occurrence and oft-repeated, consult a vet. Coughing might be due to disease of the lower respiratory tract.

'Cuddly-tame'

This expression is usually applied to young hand-reared cockatoos and macaws. It is especially appropriate for the white cockatoos (*Cacatua* species). They crave affection and love to be cuddled. Most other parrots do not enjoy the literally all-embracing contact. They only enjoy having their head scratched—but cockatoos do not object to physical contact on any part of their bodies.

I see the fact that cockatoos like to be cuddled as distorting the image of what a cockatoo really is. It is not a winged dog or a winged cat. Neither is it a baby-substitute. Or, rather, it should NOT be. Alas, sometimes it is.

The usual story is that the newly acquired cockatoo, not yet even weaned (although fully feathered) is a little love sponge. It craves affection and the new owner finds it irresistible. He or she can hardly leave it alone. Big mistake! As the novelty wears off the cuddling time is almost imperceptibly reduced. The cockatoo then starts to become demanding. It screams. And screams. The cockatoo can be pacified only by taking it out and giving it a cuddle. This is a Catch-22 situation. Don't let it happen.

When you bring home your cockatoo you should already have decided how much time, and when, you can offer your cockatoo. I would suggest that, if you go out to work, this time consists of half an hour before you leave, and a couple of hours each evening. A cockatoo will not be satisfied with less. If you don't have this length of time to spare, don't bring home a cockatoo.

D

Destructiveness

The need to gnaw is very strong in nearly all parrot species. It is one of the characteristics which set parrots apart from other birds. If you cannot allow parrots to fulfil this need, keep ducks or finches! Gnawing is essential for their wellbeing. Most parrots are not destructive in the home if this need is met. Cockatoos are an exception, since they might be described as winged demolition agents. Unlike many parrots, they excel at taking things to pieces. They can remove screws with ease and, in an aviary, staples and C-clips from welded mesh. In fact, they have been known virtually to dismantle an aviary of insufficient strength.

Macaws can also do a lot of damage, if permitted to. A macaw has been described as being like a toddler equipped with a chainsaw let loose in the house. Some people spoil their parrots to the degree that they let them carve up the house. They rip off the wallpaper, gouge holes in the walls and gnaw furniture. After a while this behaviour will not be tolerated and the parrot might be banished. There are several ways to stop it. If there is a small unused room in the house, it should be stripped of furniture and carpets and equipped with some large branches, swings and toys. A framework of welded mesh should be erected at the window—or used when the bird is in residence. It is necessary for parrots to work out their destructive urges and freshly cut branches are by far the best objects for this purpose.

Devoting a room to a parrot is out of the question for most people. Therefore Perspex should be fixed to the wall in the area where the parrot is most likely to gnaw. A stand should be made which holds a fresh-cut branch. The branch should be

renewed as often as possible. Fruit trees, elm, willow, poplar and hawthorn are among the species which can be used. If a parrot starts to gnaw furniture, it should be returned to the cage immediately, with a firm: 'Don't!'

Question
How can I stop my Goffin's Cockatoo from gnawing the top of the door when he comes out for his daily fly around?

Answer
No, the answer is not to clip his wings. The benefit he receives from this wing exercise far outweighs the destruction carried out. However, for his own safety, a cockatoo or any parrot should be discouraged from perching on the tops of doors. It would be so easy for the bird to be killed or injured if the door suddenly slammed shut due to a draught or strong wind. You

must divert his gnawing activities to another area. This is easier said than done, since full-winged cockatoos are usually too active and inquisitive to stay on a playstand for long. He must be enticed there with a range of chewable items, such as small fresh-cut branches, pine cones, offcuts of untreated wood left over from do-it-yourself sessions, leather toys and short lengths of cloth. As some parrots are nervous of a number of new items, these might need to be introduced gradually.

I would suggest that the door is kept closed when the cockatoo is free in the room so that he has to find another perch. Fix a natural branch fairly high in the room in a convenient location where a length of plastic carpet protector can be placed on the floor beneath it. You have to accept that there will be a need to protect the carpet from droppings and debris. Go to the trouble to fix this securely because it will be the focus of the room in due course. If you are wise, you will protect the immediate area of wall with a very hard material—because chewed up wallpapers and holes that need filling will not endear your cockatoo to the man about the house. If this perch is ignored, try placing one in an area which appears more attractive to him. Once the perch is accepted, which it should be if it is higher than the top of the door, be sure to maintain his interest in it by supplying a changing supply of items for destruction or eating. Whole walnuts are excellent. You might feed them only in this location. Objects such as the centres from toilet rolls are not of much use in these circumstances because they are destroyed too quickly.

Question
My Amazon is generally very well behaved during his periods of freedom in my living room. The only problem is that he focuses his attention on a fabric-covered sofa which he always wants to bite at. How can I prevent him damaging it?

Answer
Try to persuade him to sit on a stand or play gym with fresh-cut branches for gnawing. Failing this, cover the sofa when he is out. Note, however, that sofas upholstered in leather are much less attractive to parrots. They do not like the cold, slippery surface. Ensure that he has a plentiful supply of toys and items to

gnaw in his cage, so that he will be in a less destructive mood when let out. Also heed the heart-breaking story of a Green-winged Macaw. He was allowed to destroy a sofa while his owner was on holiday. He ingested some of the interior fibres with fatal results.

Question
Why does my Blue and Yellow Macaw destroy the wooden perches in her cage within days? I cannot find any wood strong enough to prevent this. Can I use metal perches?

Answer
To a large macaw, using the beak to destroy is a natural activity which can never be altered. She will pay less attention to the perches if she has a regular supply of pieces of thick branches from trees of harder wood. You might contact a tree surgeon with the request to keep for you suitable pieces of wood or branches which you can saw up for gnawing and for perching. On no account use metal perches. They provide a very harsh surface for the feet—cold and hard. (In outdoor aviaries they are very dangerous. A parrot could lose its toes if it roosted on a frost-covered metal perch.)

Dominance

Elsewhere (*see* Aggression, *see* Training) the reasons why parrots attempt to dominate people, and how these can be corrected, are described. Parrots can be so subtle in achieving dominance over a person, that at first the human hardly realises what is happening.

One form of dominance might be described as height dominance. In a parrot's mind, height equals superiority. If he is consistently at a higher elevation than your eye level, he will perceive himself as dominant over you. His cage should therefore be situated where this does not occur. When free in the room, he will be much more difficult to control and to return to the cage if his favourite perching place is a high one. Try to discourage perching on pelmets, for this reason. A play stand or play gym is best placed on a low table, such as a coffee table. Many parrots love rope swings which are suspended from the ceiling; make sure that the rope or chain holding the swing is long so that the swing is not out of your reach. Despite these precautions, some parrots will still favour a high perching place. In this case they should be trained to step on to a small wooden ladder which brings them within easy reach.

If a parrot at a high elevation perceives itself as dominant, the reverse applies. This is why I hate the use of stacking cages for parrots and for other birds. In fact I see it as a form of cruelty because the bird on the lowest level will feel so inferior its quality of life will be destroyed. One lady told me how due to lack of space she had bought three stacking cages. One of her Greys was at the top, a pair of conures were in the second cage and a young Grey was in the bottom cage. On my advice she moved them. It seemed unlikely that the young Grey would be able to develop its personality or to learn to talk in such a lowly position. In addition, constantly to see the lower half of people passing by is to condemn a parrot to a life totally lacking in stimulation. The moral here is clear: if there is not enough space in the house to house parrots at human eye level, another parrot should not be obtained.

E

Euthanasia

When age or incurable illness means that the quality of life of a much loved pet has deteriorated to an unacceptable degree, euthanasia must be considered, then carried out by a vet. The decision is hard to make—but delaying it results in prolonged suffering to the bird. A humane and rapid end is the least we can do for a companion who has given so much pleasure.

Eye contact

Eye contact with a human can appear very threatening to a nervous parrot. Avoid this. It can also initiate aggression in combative birds. If I have to enter the aviary of nervous or aggressive parrots I always avoid eye contact or even looking in the parrots' direction. In this way my presence is better tolerated. (Crouching down also makes a person appear smaller and less threatening.) Note that a nervous bird does not blink while looking at a person or other object of fear.

Another action in which eye contact plays a large part is approaching very close to a cage containing a parrot and staring head-on into its face. While some parrots can accept this, others, especially some Greys, hate it. When approaching the cage of a bird who does not know you, stand at least three feet (about a metre) distant.

One of the methods for reprimanding a bird recommended by some American behaviourists is known as 'the evil eye'. When a parrot does something wrong, you glare at it and fix that stare for a few seconds. Personally, I would never use this method. First, glaring in humans evokes unfriendly and antagonistic behaviour. I would not want to put myself in this frame of mind towards my parrot. Secondly, all of the instructions we give to our parrots are vocal or accompanied by words. Glaring would therefore seem to be less effective for most species. I am told that it works well with some Grey Parrots. I believe that a more appropriate reprimand is the short sharp: 'No!' or 'Don't', at the same time suddenly pointing the index finger to emphasise the word.

F

Fear

Few emotions have such a profound effect on behaviour as fear. How can a human learn to recognise this emotion in parrots? There are two methods: watching their behaviour and the expressions in their eyes. The latter suggestion might be met with disbelief in some quarters, which would not surprise me in the least. Humans believe that they rely greatly on the expressions in another person's eyes to interpret their emotions. In actual fact, they rely equally on facial expressions. Birds' faces are covered by their feathers. In some birds ruffling of the head feathers or erecting the crest (in cockatoos) might express fear, anger or aggression. But this is not the case with others. An experienced human observer can detect a change in the expression of the eyes of birds with which they are familiar. The easiest emotions to detect are fear and curiosity.

Not long ago I removed from a magazine a large poster-type photograph of one of my favourite species of small lorikeet (a nectar-feeding parrot). The photography was superb and the lorikeets' feather condition was perfect. I placed the picture on my office wall. But I soon took it down. The pair had obviously been photographed in a photo cage in which they had just been placed. The fear in their eyes at finding themselves in a strange place outweighed the beauty of the photograph. Parrots' eyes show anxiety very clearly to the person who takes the trouble to look.

There are other signs of fear in birds. The plumage is held tight to the body, making the bird look smaller. The eyes stare unblinkingly as though in the split second of a blink a predator might descend and grab the bird. There are no signs of relaxation (see pages 34-35), such as stretching, preening and dozing. The bird may emit an alarm call or it may be too frightened to utter a sound. Some parrot species will even 'freeze' in the middle of a movement, the foot poised motionless above the perch. It is as though the slightest move might be detected by a predator.

Always remember that parrots and other birds behave in a way which allows them to take flight instantly if they believe they are in danger. In Australia, one can see small lorikeets, such as the Musk and the Little, hanging head-down from their nest entrance. In this position they are poised ready for flight and can take off in a split second. In captive birds it is a different story. So many parrots have had their wings clipped that this fundamental behaviour is denied them. I suspect that this has more serious behavioural implications than most of us realise. Birds have been programmed for millennia to flee from danger. They know instinctively how to escape from a predator but these birds with clipped wings are unable to do so. What conflict does this produce within them? We shall never know. The stress for a wild-caught bird might be serious enough to induce feather plucking, for example.

Flight-intention movements

Flight-intention movements can be observed in many wing-clipped parrots. They are often the result of fear but sometimes they are just an indication of a desire to be moved to a different place. For example, I saw flight-intention movements in my Amazon (too old to fly), whenever she wished to be moved from her stand to her cage. Thus one must learn to interpret this behaviour.

Flight-intention is indicated by a crouching posture, with the parrot's head held forward, the wings quivering and tail slightly raised. It is often accompanied by quiet vocalisations of a kind which will not be heard at any other time. If the parrot is unable to fly, this behaviour will be repeated several times. I have a Rajah Lory which is flightless due to an old wing injury. Her flight-intention movements are accompanied by a rapid bobbing of the head.

If a companion parrot is in a constant state of fear it is

unlikely to become tame or to learn to talk. Fear can be induced by the close confinement of a small cage because escape is impossible. Some fearful birds will improve when placed in an aviary and, if they can fly, the flight-intention movements will no longer be observed.

Question
I recently bought a pair of Orange-winged Amazons. They are very nervous and have started to bite off their tail feathers. When I go near them one of them will raise its foot. Why does it do this?

Answer
The stance of one-leg-raised is typical of an Amazon parrot which is experiencing fear. It is, in effect, trying to ward off something of which it is afraid. Unfortunately, your birds are wild-caught adults and may take a very long time to settle down in captivity. They are still suffering from the stress of capture. This is almost certainly why they are biting off their tail feathers. On no account remove the stumps as pulling out tail feathers often creates problems. The feathers should be moulted normally in late summer or autumn. Hopefully by then they will have adjusted better to captivity. This is more likely to happen if you place them in a large outdoor aviary. At present it is the close proximity to humans which is making them so nervous.

Question
I recently bought a Grey Parrot. I was told that it was tame but when I got it home I found that it growls whenever I go near it. How can I tame it?

Answer
Growling in Grey Parrots is a sign of extreme fear. It would be unusual for a captive-bred hand-reared bird to do that. You have almost certainly bought a wild-caught Grey. If its eyes are yellow, indicating that it is an adult, it will be very difficult or impossible to tame. Taking adult birds from the wild is a very cruel practice as they live in a permanent state of fear and stress, afraid of everything. Most captive-bred birds wear closed rings. People should refuse to buy birds which do not have closed

rings. In purchasing the cheaper (half the price) imported birds, they are encouraging a cruel and unacceptable trade. It is not unknown for the few wild-caught birds which survive (the death rate is appallingly high) to become calmer but it may take many months and a very patient and sympathetic person to achieve this. Most wild-caught birds are suitable only for an aviary.

Feather plucking

Feather plucking is an outward sign that there is something wrong in a parrot's life. The hard part is deducing whether the cause is psychological, due to ill health, a dietary problem or due to something in the immediate environment. I am constantly amazed at how many people believe that feather plucking is caused by the presence of mites or lice. This is hardly ever or never the case.

Certain groups of parrots are much more susceptible than others to plucking out their own feathers. These are Greys, cockatoos and Eclectus Parrots. In a large proportion of these cases, the cause is psychological. Stressful events are likely to trigger feather plucking. These include the absence of the person to whom the bird is very close, a change of home, even being moved to a new location within an established home, the sudden introduction of another parrot or animal (even a human baby) of whom the parrot is jealous, or being placed with another parrot who is dominant or even aggressive. Permanent stress can also cause feather plucking. Possible causes are teasing by a member of the household, and being kept in an open cage in an area which lacks seclusion, or very low down so that the bird cannot see what is going on. Bearing in mind a parrot's need to gnaw, being kept in a wire cage, on a plastic perch and never given anything to gnaw or play with could initially result in over-preening which develops into feather-plucking.

Ill health is a common cause of feather plucking. This is why it is recommended to take a plucking bird to an avian vet soon after the habit commences, if there is no obvious reason. The fact that the bird looks well and behaves normally otherwise does not rule out the fact that it is sick. To try to establish whether illness is causing a parrot to pluck itself, a competent avian vet will carry out a complete physical examination, faecal analysis, bacterial cultures, blood analysis and, if available, testing for viruses. Note that some viral diseases, such as PBFD (psittacine beak and feather disease caused by the circovirus) and polyoma cause feather loss. This might be confused with feather plucking by the uninitiated. A parrot usually starts plucking itself on the breast. Obviously it cannot reach the feathers of the head; if these are missing the most likely cause is PBFD or, if the parrot has a psittacine companion, the other bird could be plucking it. This is not uncommon and usually starts with the head feathers.

Pain can result in a parrot plucking itself in a certain area, for example, near the liver if it has a fatty liver. A skin irritation is another cause. This is why it is so important to ensure that a parrot is sprayed or bathed on a regular basis, except when it is very young. An infestation of *Giardia*, a protozoal parasite, is a common cause of feather plucking in Cockatiels, and can affect other species. A bacterial infection of the skin or feather follicles could cause the bird to pluck itself.

A bad wing-clip can cause parrots to bite at their wing feathers. This is especially the case if the feathers are cut using blunt scissors which crush them. One unfortunate young Grey had its life ruined by a bad clip which resulted in it being in constant pain and discomfort. New feathers broke off before reaching half their length, leaving a 5 cm (two-inch) shaft. The surrounding skin was swollen and painful. The clip probably caused feather follicle damage resulting in the new feathers being deformed. This might be a temporary or permanent condition. If the damaged feathers were not moulted they might have to be removed under anaesthetic.

Dietary deficiencies (of Vitamin A, minerals or amino acids) or excesses (over-stimulating items) and food allergies (various foods including certain seeds or seed dust) might also cause feather plucking. Zinc or other heavy metal poisoning could also be to blame. A parrot might be absorbing small quantities of zinc which gradually build up over a period of time. This might be from some object with which the parrot has daily contact, such as a food container, a swing made of galvanised wire or some part of the cage. This cause is usually overlooked.

Some female parrots pluck themselves when they come into

breeding condition, regardless of whether they have a mate. Usually only the breast feathers are denuded, and these grow back after several months. But this behaviour may be an annual occurrence, which starts at approximately the same time each year. In the UK May is the month when females of spring and summer breeders are most likely to be affected. There is a possibility that fleas from dogs or cats could bite parrots or even hide under a wing, causing great irritation. This could result in the affected area being plucked. If you have animals in the house, make sure that they are treated for fleas on a regular basis.

A lack of direction, especially in an adolescent parrot, could result in feather plucking. In the wild, parrots would learn a lot from their parents and other mature birds. They learn how to behave. In captivity, young birds can become very confused by lack of leadership. They need someone to guide them. Teaching them to step up and to respond to other simple commands creates an atmosphere of discipline in which they will thrive. This could prevent or even curtail feather plucking.

Stopping plucking

In many parrots feather plucking becomes a habit which is very difficult or impossible to break. This is why steps to stop it must be taken as soon as it starts. Suggestions are as follows:

- When you see your parrot plucking never yell and shout at it to stop as it may perceive feather plucking as a means of gaining your attention. Although I do not recommend punishment as a means of altering behaviour, the mild form of punishment involved when the parrot's favourite person leaves the room, can be effective. Greg Glendell related how he quickly trained a neglected Grey Parrot called Freddie to stop plucking in his presence. The bird was on a temporary visit to him while he tried to resolve some of its problems, and he was enjoying Greg's companionship. When he plucked out a feather while sitting on Greg's hand, Greg said 'No!' firmly, put Freddie down and left the room for ten minutes. Freddie called to Greg, who refused to answer. Greg returned and pretended Freddie was not there, until

Freddie crept up to him to have his head scratched. About half an hour later Freddie pulled out another feather—so Greg immediately put him down and left the room. The fourth time he pulled a feather was the last. He seemed to have learned that if he did this in company, he would be left on his own. This training session did stop him plucking when other people (apart from Greg) were present (Glendell, 1998b).

- Carefully examine your parrot's recent history. Make a note of any kind of change in routine, cage location, diet, family members or visitors and in the cage or its contents. (The addition of a swing to a conure's cage caused it to scream and to pluck, *see* p 37.) Note whether any birds have been added to the household or moved nearer the parrot, or whether any other pet is making the parrot nervous. Consider whether any aerosols, household cleaners, including carpet cleaners, have been in recent use. If so, ban them from the room the parrot inhabits. If you need to consult a vet, take these notes with you.

- Ensure that the environment is not too dry and that the bird has a bath or shower at least twice a week.

- Provide fresh-cut washed pieces of branches from apple, willow, ash, elm, hazel, hawthorn, poplar, etc, at least once a week, plus toys and ropes for gnawing. The round circular woven rope swings are excellent for this purpose, except for the most destructive large macaws and cockatoos.

- Provide a balanced diet which includes fresh fruit and vegetables.

- Consult an avian veterinarian or a homoeopathic vet. (Contact the British Association of Homoeopathic Veterinary Surgeons.) A three-year-old female Eclectus who was shy, underweight and plucked, was first given a homoeopathic remedy Silicea C30, then given Bach flower remedies, colour therapy and acupressure. After only two weeks the female Eclectus was described as self-confident and happy and, furthermore, she had stopped plucking. (Wagner and Sonnenschmidt, 1997.)

- Do everything possible to provide more stimulation. Constant human companionship and moving the parrot from one room to another during the day, so that it is not continually looking at the same objects, is recommended. If it must be left alone, leave the radio playing music, or position the cage where it can watch the type of television programmes which usually attract a parrot's interest—wildlife films, pop music and cartoons. Television is recommended for short periods only.

- Provide food items that will keep a parrot occupied, such as whole walnuts, especially for species for which they are a challenge to open, or pine nuts for smaller species. Or put these items in very small cardboard boxes so that they have to work to find them.
- Ensure that the diet is not deficient in Vitamin A. So many parrots suffer from this deficiency and it causes a multitude of problems.
- Consider providing a much larger cage. Cages which are too small can be very stressful for many parrots, especially those which are nervous. Such a cage might be made from welded mesh instead of buying an expensive cage.

Feather plucking can take several forms, from removing whole feathers (the most common occurrence) to just stripping away parts of them. One vet believes that it may be addictive because plucking out a feather causes the release of endorphins. These are natural opiates which perhaps give the plucking bird a 'high'. The answer might be to use an endorphin blocker such as Naltrexone.

Owners of plucked birds should note that if and when the feathers start to grow again, they may look very untidy and grow at strange angles. It could be one to two years before the plumage appears to be in reasonable condition—and longer than that before it is perfect.

Question

I have a Nanday Conure, four years old, which started to pluck himself just over one year ago. He has removed all the feathers he can reach. He has a varied diet, my company all day long and is a lovely pet who never bites. I have tried Bach flower remedies without success. He wore a plastic collar for some time to stop him plucking but after taking it off he was virtually naked within a week. What else can I do?

Answer
Consult a good avian vet in order to have the relevant tests carried out. Even though your bird is young, you cannot rule out disease. If the test results show your conure is in good health, consider all the possibilities listed above. In my opinion, using a collar is very useful to prevent a parrot from biting at, for example, a wound which is healing, but it does nothing to address the underlying cause of feather plucking. A collar causes a lot of stress to most birds and, as you have discovered, it is only a temporary measure to stop feather plucking. It would be pointless to use it again. You do not state how long you used the Bach flower remedies. Perhaps it was not long enough or the appropriate remedy was not used.

As your conure was three years old when he or she started to pluck—the age of sexual maturity—there is a strong possibility that the frustrated desire to breed is causing the problem. You might consider having your bird DNA feather sexed, then obtaining a mate for him or her. Conures are among the most sociable of all parrots and an aviary, even an indoor one, and a mate could enormously improve the quality of life for your Nanday. Bear in mind, however, that even if he or she stopped plucking, the feathers might not grow again in all areas (on the breast, for example), because by now the feather follicles might be destroyed. If you decide to obtain another conure, I would suggest that you quarantine it away from your bird for five weeks or so. The quarantine is best not carried out in your house because if the two birds can hear each other they will call backwards and forth—and the noise level will not be pleasant.

Feeding
For parrots, feeding is a very important part of the day's activity—not only for the obvious reason that it provides nutrition. In the wild, feeding is a social activity, performed simultaneously by family groups or small flocks, or large flocks. For captive parrots, feeding times are even more important because they break up the monotony of the day—just as they do for humans. This is why I believe that whether in the home or the aviary, parrots should be fed twice a day. A parrot kept in a room where people are eating will usually go down to the food dish at meal times. If the food dish is empty he will scream. In fact, offering some

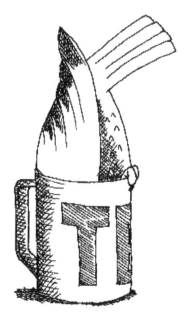

Parrots love to forage

food from the table will do no harm if sensible items are chosen. Birds fed on seed-based diets will actually benefit from this practice, as foods such as vegetables (cooked or raw), pasta, lean meat and cheese, will introduce nutrients which will almost certainly be missing from a diet of seed and fruit.

Parrots have more taste buds than any other bird—in the region of 300 to 400. They can thus discern the most subtle flavours. Most birds have between 100 and 200 taste buds (Rensch and Neunzig, 1925). For example, given a piece of orange which is not sweet, most parrots will ignore it. The owner might believe that their birds have tired of oranges—but offer a sweet one and it will almost certainly be eaten. So if your parrot suddenly refuses an item which is normally relished, taste it. On the other hand, some birds will eat a certain item eagerly, then lose interest in this food for a while. Thus variety in the diet is important.

Food can be used to keep parrots occupied and stimulated. There can never be enough sources of occupation, especially for the larger species. Zoos are becoming increasingly interested in 'work to feed' devices, like feeder puzzles. As an example of one which is easily constructed, a plastic electrical wire reel with a hollow centre can be transformed into a rolling dispenser for cockatoos and macaws. Holes can be drilled into the central tube (using a quarter-inch bit on a power drill). PVC plumbing fittings block the ends. One end is secured with a glue gun and the other end is refillable. The parrot must roll the reel about the floor to dispense food, usually one morsel at a time, as it falls through the quarter-inch openings.

Although it is normally wise to quarantine new arrivals away from other birds, the close presence of another parrot will usually persuade the newcomer to start feeding. Eating is an activity which is engaged in by all flock members simultaneously. Therefore, another parrot in a nearby cage is the best stimulus. Also bear in mind that if a parrot has, for example, fed from a brown plastic food cup for all or most of its life, it might be reluctant to sample food from a stainless steel container. Try to find out what type and colour of food container it is used to.

Question
When I give my parrot toast why does he deliberately drop it in his water pot before eating it?

Answer
For the same reason as some people dunk biscuits in their tea! He prefers it soft.

G

Grieving

There is no doubt that parrots can grieve for the loss of a human or psittacine companion. This can affect their behaviour. Their appetite may suffer and they become quiet and withdrawn. This is what happened in the case of a Grey Parrot—a youngster which I had hand-reared. She went to live with a friend to whom I had given an Eclectus Parrot. Later, a male Grey joined the household—but the two Greys were never friends. The female Grey and the Eclectus were firm friends and spent hours chatting to each other. When the Eclectus died, 13 years later, the female Grey stopped talking. It seemed that she was still grieving, months later.

H

Handling

A few parrots will go to anyone—at least while young—but most will not. Other members of the family should interact with the parrot from the outset, if they wish to handle it. If it is a young bird, they should remove it from the cage once daily (at least) and perhaps walk around the house with it, showing it things. Just offering a favourite tit-bit, perhaps through the cage, is not enough to form a bond. This attention is most effective in a room away from the parrot's cage, especially one which the parrot does not know well.

Should a parrot be handled by non-family members? I believe that it is unreasonable to expect a parrot which allows itself to be handled by various people to go to someone it has never seen before. If that person is nervous or inexperienced in handling a parrot, there is a danger of being bitten. When a parrot starts to know someone who has become a regular visitor to the household, handling might be tolerated. But if the visitor is someone who is close to the parrot's primary human companion (or worse, only human companion), an element of jealousy might make handling impossible or unsafe. There is, though, the other type of parrot, especially the cockatoo, which is very friendly and ever-demanding of attention, who will enjoy the opportunity to interact with anyone interested. In fact, such handling is likely to prevent attention-seeking behaviour such as screaming. With more cautious species, such as Greys, unless they are very young, they should not be expected to go to other people on demand. (*See also* Training.)

Hands

Many parrot owners do not realise that some parrots are afraid of hands. When initially approaching a nervous or wild-caught parrot, put your hands behind your back or in your pockets so that the potentially most threatening part of your anatomy is hidden.

The new owner of a parrot may be very disappointed to discover that his or her bird looks with great suspicion at hands and backs away. What can be done? When the parrot is inside his cage, talk to him often, always with your hands behind your back. If he is interested and comes closer, and seems to want contact, try offering a favourite item of food, so that hands come to be associated with something good. Talk to the parrot while holding the tit-bit for a couple of minutes. Always move your hands slowly.

For a parrot which has yet to overcome its fear of hands, teaching it to 'Step up!' on to a hand will be out of the question. He might be taught to step on to a tiny ladder or a piece of dowel, instead of the hand, so that there is some control over the parrot's movements. The bird (full-winged) can be allowed to perch on a seated person's knee.

Case history

Beak contact is common between parrots which are very familiar with each other—mates, siblings and other close flock members. Some parrots seem to regard the human nose as the equivalent of a beak. They will accept nose contact but not hand contact. This was the case with a male Citron-crested Cockatoo which came into my care. Like many wild-caught parrots, he was afraid of hands. As long as my hands were behind my back I could approach him very closely. He seemed to seek contact, coming right up to the edge of the cage. What I did next is not something that I would recommend, except to people who can read parrot behaviour well. Cockatoos love to have their foreheads rubbed, so I rubbed his with my nose. It was taking a risk, because the cockatoo could have inflicted a very unpleasant bite. However, I was confident that he would not do so. A few days later I was able to bring my finger slowly up to my nose while rubbing his head and scratch his forehead with my finger. In this way, the cockatoo lost his fear of hands. I repeat that I do not recommend this to other people and certainly not with a Grey Parrot. Perhaps it would only be effective with a cockatoo. I relate this case only to demonstrate that the fear of hands can, in some cases, be overcome in a short space of time if carried out in a sensitive manner.

'House-training'

It was difficult to know which term to use here—but since everyone understands the expression as relating to dogs, it seems appropriate.

Question
Can I teach my Cockatiel not to make a mess over me or the furniture when she is out of her cage?

Answer
Yes, you can—if you are observant. Before a bird defecates, it lifts its tail slightly or alters its body posture in a way which you will soon recognise. When you see your Cockatiel do this, speedily take her on your hand and put her on her stand, back in her cage or hold her over a sheet of newspaper. You must decide where before you start training her. When she produces her droppings, give a command. Keep to the same one. Then praise her. She will quickly learn the routine. The rest is up to you. When she is out, make sure that every 30 to 45 minutes you give her the opportunity to produce her droppings in the appropriate place. Watch her body actions, as these will indicate when she wants to go. A small bird like a parrotlet or Budgerigar will need to go every 15 to 30 minutes. A larger species such as an Amazon or a macaw can wait more than one and a half hours. Sleeping parrots can go for hours without needing to empty their bowels. To 'house-train' a parrot is usually quite easy.

Intelligence

Those who work closely with the larger parrots know that they are among the most intelligent of all birds. This intelligence may be matched or surpassed only by members of the crow family. However, it is difficult to measure intellect. A good indication is given by the size and weight of the brains of various species. Of those which have been studied, the intracerebral weight index of a Blue and Yellow Macaw (*Ara ararauna*) is 28—the highest level among birds. That of a domestic chicken is 2.9 and some raptors reach 8.3.

Why do some parrot species show a greater degree of intelligence? Scientists believe there are probably two determining factors: the foraging behaviour (how cognitively demanding it is to obtain food) and the social complexity of that species. If this is true, it helps to explain why the most intelligent species are the most susceptible to feather plucking unless they have foraging enrichment.

In recent years a number of scientific experiments have been carried out that claim to assess the brain power of parrots. The researchers usually conclude that this is equal to that of a four-year-old child. However, such comparisons are not really appropriate.

In 2013 Alice Auersperg worked with captive-bred Goffin's Cockatoos at the University of Vienna. Now anyone who knows this species is aware that it is very clever—but what Dr Auersperg found was a type of decision-making that is rarely found except among humans and a few large-brained animals. The cockatoos were able to refuse an immediate nut for a better one for more than one minute. This mirrored an experiment made during the 1970s in the USA when small children were given an edible treat. They could either eat it immediately or wait fifteen minutes and get two treats. The cockatoos were offered a pecan nut and actually held it in the beak but received a cashew nut, which they preferred, if they waited up to 80 seconds and did not eat the pecan. The cockatoos clearly could withstand temptation and understood the advantage of future benefits over immediate ones! Remarkable!

Some parrots, such as Greys, have a reasoning power that is quite extraordinary. The famous Grey Parrot, Alex, was for 30 years the "student" of animal psychologist Irene Pepperberg. She claimed that his intelligence was on a par with that of dolphins and great apes, that he had the intelligence of a five-year-old human and the emotional intelligence of a two-year-old human.

Because Grey Parrots are such good mimics and because some people who look after them have the sense to teach them the names of objects, many people are aware that parrots have the potential to use words creatively – surely an indication of a superior brain – rather than merely mimic words. For example, parrots can learn to ask for what they want – in a language so far removed from their own. Those who live closely with a parrot often observe actions that indicates exceptional cognitive and reasoning abilities.

Intelligence might be defined as including problem-solving, learning new skills (even tool-making) and emotional intelligence. Just as in human beings, the level of intelligence in parrots varies widely according to their environment, their genes and their life experiences. However, what we perceive as intelligence often relates to the characteristics of the species, each one of which has evolved in a slightly different environment, thus acquiring different skills to differing degrees. For example, the Kea is often considered to be the most intelligent of all parrots. It evolved in a harsh mountain environment and has had to learn a multitude of skills in order to survive.

J

Jealousy

Many people question that this emotion can be felt by a 'mere bird'. I have no doubt that this is the case. A parrot in a single-bird household is, not unnaturally, very jealous if a second parrot is acquired. This is something which should not be undertaken without a great deal of thought. Your relationship with your original parrot could be ruined for ever; or the new-comer could trigger off problems which are difficult to deal with, such as feather plucking or screaming.

But if the decision to go ahead is made, what can be done to soften the blow? A lot—if you are prepared to go to some trouble, as did Kirsten White. She suggested that her Moluccan Cockatoo would feel the same way about the young Grey Parrot she was proposing to add to the household as she would if her husband had come home and announced that he had acquired a second wife. She therefore set up the new cage in advance and, at the same time, gave the cockatoo a number of new toys. She then set about making a Grey Parrot doll using grey socks and a red sock. The head consisted of a photo of a Grey's head cut from a bird magazine. The effigy was then attached to the perch in the cage, with the side of the face with the picture facing the cockatoo. From then on, for the next three weeks, Kirsten treated the parrot doll as though it were real—talking to it, taking it out, cleaning the cage and playing with its toys. At first the cockatoo yelled and displayed at the intruder, but its impassivity slowly turned the cockatoo's aggression into curiosity. After a while he began to talk to it, instead of screaming at it. In due course the young Grey arrived. The cockatoo was rather surprised when it climbed to a new perch but he soon

accepted her. This was no doubt because he had already decided that the cage's occupant was not worth making a fuss about. The gradual introduction strategy had succeeded beyond all expectations (White, 1999).

Be aware that parrots can become intensely jealous of other new pets, such as a puppy. This is not surprising when you think of all the attention a new puppy receives. It is instantly the focus for everyone who enters the house. For a parrot which has been used to hogging the limelight, this is not easy to tolerate. Try to keep the puppy in a different room to the parrot for most of the day or take it there when visitors want to play with it.

Question
My two-year-old Sun Conure, Max, has become unbearably noisy lately. I am considering buying another conure to keep him company as, hopefully, this would quieten him down. However, I fear that his reaction might be one of jealousy. What would you advise?

Answer

Unlike some birds which are intensely jealous of a new arrival, conures usually make friends with one of their own kind very quickly, as they crave the companionship of their own species. There is a good chance that buying a mate for your assumed male would quieten him down. It would be pointless to buy another conure of the same genus but of a different species, as the two birds would probably want to breed, and it is unwise and unethical to produce hybrids. You also need to make up your mind whether breeding is what you want your conure to do. If not, do not buy another bird unless you feel you really cannot tolerate his noisiness any longer. While two conures can also be very noisy at times, they are more likely to be occupied with each other and would yell mainly if alarmed. Max might remain tame but he would probably become aggressive, if he is a male, when breeding.

If you buy another conure, it must first be kept in a separate cage as Max would probably attack the newcomer if it was placed in his cage. If the two seemed compatible, they should be placed in a large cage which neither has previously inhabited. However, before buying another conure, you need to establish beyond doubt Max's sex. All you need to do is to pluck several feathers from his breast and send them to one of the avian diagnostic companies which advertise in bird magazines. At the time of publication the fee was less than £20 per bird for DNA sexing. Moulted feathers should not be used for this purpose—only freshly removed ones.

K

Kissing

Question
My Cockatiel loves to sit on my shoulder and give me kisses. But if I open my mouth he puts his head inside and nibbles at my teeth. Why does he do that? And could this be harmful? Someone told me that I could get an infection from him.

Answer
Your Cockatiel is not alone in this strange habit. Many parrots will do this, if permitted. The reason is unknown but it might be that they like human saliva. There is a very remote possibilty that you could get an infection from him, but the chances are much greater that you could transfer an infection to him. This is why kissing your Cockatiel or any other kind of bird is not recommended, especially if you have a sore throat. The streptococci bacteria which are often associated with this condition can be transmitted to birds. The same is true of at least some human influenza viruses.

L

Lameness

The usual advice to anyone choosing a young parrot is to ensure that it is healthy—active, bright-eyed, with no discharge from the nostrils or matted feathers around the face that might indicate disease. However, a vitally important aspect that is often overlooked is that its bones are healthy. It should be standing well, gripping the perch and moving about easily, have a straight spine and be able to fly. Post-weaning, it should not have a stance which makes it low to the perch. If it has, suspect a calcium deficiency, or a calcium/phosphorous imbalance in the rearing diet, resulting in metabolic bone disease. This is often known as rickets. Or the bird might have received calcium but not in a form with Vitamin D^3 included or without access to sunlight. Both aid calcium absorption. The problem is common in parrots fed an all-seed diet without calcium/vitamin D^3 supplementation.

Unfortunately, many breeders are producing parrots, from the smallest species to macaws, that are not just lame, but deformed. Breeders may not realise this because it is not always evident to the inexperienced eye but is very apparent when an x-ray is made.

Such young birds have rubbery bones that result in painful fractures. This is usually most apparent in the legs and feet. It is a tragedy because the bird will either be deformed for life or it will have to be euthanised. In adult birds suffering from metabolic bone disease the skeleton gradually weakens, also resulting in fractures or breeding failure.

Case history

A six-month-old Cockatiel was taken to a vet because lameness in its right leg had been noticed since it was acquired four weeks previously. It was very thin and could not grip with either foot. An x-ray revealed that the leg was fractured, the spine was bent and the bones were poorly mineralised. The diagnosis was metabolic bone disease. Emergency intensive care was started with the Cockatiel being placed in an incubator at 26°C. If placed in a cage it would have climbed and possibly broken more of its brittle bones. It was tube-fed a hand-rearing food with added calcium, vitamins, minerals and amino acids. Pain relief and antibiotics were also given. Recovery was slow because it takes months to re-mineralise bones.

After four days the Cockatiel was taken home. The owners had purchased a brooder where it was kept for a recommended six to eight weeks. The diet was changed from seed only to millet, dark green vegetables (beet, spinach and kale), organic pellets and a calcium/Vitamin D^3/magnesium supplement. Two weeks after the first treatment the Cockatiel had gained 10g and was starting to support weight on its legs and feet (Sacks, 2014).

This was a lucky bird in that the owners were prepared to spend a considerable amount on veterinary care and to follow the advice given. Bone disease should never happen. Some breeders need to be better informed and more conscientious. The pain and deformity that birds such as this Cockatiel suffer is due to ignorance – and is inexcusable.

Laughing

Amazon parrots—above all other species—have a marked inclination for mimicking laughter. They seem to find the sound stimulating and can mimic exactly the laugh of an individual. The sound of laughter on radio and television can also cause them to mimic the sound. I have never heard a cockatoo mimic or respond to laughter. Some Grey Parrots do so.

Laying

Egg-laying by single pet birds can become a serious problem. Many Budgerigars and Cockatiels become chronic layers—and this can seriously damage their health and even lead to their death. The problem is made worse by some carers of such birds who remove the eggs as they are laid. This merely stimulates the female to lay even more. It is advisable to let the female incubate the first clutch in the bottom of the cage, until she tires of this. A Budgerigar or Cockatiel will incubate for approximately three weeks and larger species for about four weeks.

If only one or possibly two clutches are laid in a year, this should not be detrimental to the female's health, provided that she has a good diet and/or calcium supplementation. But if she is a serial layer, even with a regular calcium supplement, she could start to draw on the calcium reserves in her bones, because blood calcium has been depleted. If this happens, she could become paralysed. Additionally, she would be susceptible to egg-binding and other reproductive disorders (even a prolapse of the reproductive organs), muscle weakness and infections. It might be advisable to convert some species (not Budgerigars) to a pelleted diet, as this will reduce the likelihood of calcium deficiency and dietary imbalance. Incidentally, switching some breeding pairs to pellets can stop them laying for the best part of a year. This happened to the rare species belonging to a friend who used the most expensive and vet-promoted pellets on the market.

Question

My three-year-old Budgerigar keeps laying eggs, despite the fact that we were advised to let her sit on them. After losing interest in one clutch, she will lay again about ten days later. So far this year she has had four clutches. I am very concerned that she might become egg-bound. How can I stop her laying? She used to be such a lovely pet but now she seems to have lost interest in us and only wants to sit on her eggs.

Answer

You are right to be concerned. First of all, give her a calcium supplement. The latter will normally include Vitamin D3 which aids the absorption of calcium. If your bird is tame enough to handle, try to give her a liquid supplement direct into the beak, such as Calcibor (40 per cent calcium borogluconate) obtainable from a vet. If not, obtain the glucose-based powdered supplement Nekton MSA and add this to a favourite food. As birds from arid environments, such as Budgies and Cockatiels, drink little water, it is usually a waste of time putting supplements in the drinking water. Calcium supplementation is very important for birds kept indoors; for those outdoors, exposure to sunlight aids in calcium absorption but on its own is not sufficient for many egg-laying females.

Secondly, you have to try to induce her out of laying mode. If she is laying the eggs in any kind of container on the cage floor, remove it. You might even consider putting a false base of welded mesh in the bottom of the cage, so that she cannot scratch around and prepare a nest there. Make the area where she usually lays as unattractive as possible. Next, reduce the hours of light to which she is exposed. Either cover the cage so that she has no more than 12 hours or, in the evening, move her cage to a darkened room. Also, move her cage away from its normal position or even put her in a different cage, so that the surroundings are unfamiliar. This alone can get her out of the laying habit. Provide her with new toys which keep her busy. Try to give her more attention on a permanent basis but don't stroke her as this can be a sexual stimulant which induces egg-laying. If someone could be with her during all her waking hours this would help to divert her attention from wanting to lay. If these actions are not successful, you should consult an avian vet. Hormone treatment can put a stop to egg-laying.

Loneliness

My strongly felt opinion may not be popular, but I feel that it is selfish indulgence on the part of a person who is out working during the day to buy a single parrot, Cockatiel or even Budgerigar. Birds are so different from cats and dogs in this respect. These animals will sleep when they are alone. Traditionally cats, and even dogs, inhabit unoccupied houses during the day, and this tradition has thoughtlessly been extended to parrots, in many cases. Birds need a stimulating environment. They are naturally much more active and vocal than mammals. They may sleep or doze briefly only during the middle part of the day—unless they are old or very young, in which case they might sleep for longer periods.

Question

I have a hand-reared Blue-fronted Amazon, now nearly two years old. When I get in from work every evening he screams and screams, so that I have to take him out before I can do anything else. How can I stop him from screaming at this time?

Answer

Only by not going to work! You do not say so but I surmise that your Amazon is left alone during the day. When you come in he is calling to you as he would call to his mate, if he had been separated for some hours. This behaviour is perfectly natural. For an intensely social creature like an Amazon to be left alone for hours each day is unnatural. I would strongly advise anyone who is out of the house for eight or so hours a day, five days a week, not to obtain a single large parrot. For it to suffer loneliness and boredom for the major part of its waking hours is totally unfair. In these circumstances, a pair of a smaller species of parrots would be more appropriate. Do not think that the answer is now to buy a mate or a companion for your parrot. What he needs is not what he perceives as competition, but fewer hours alone. If you work near enough to come home at lunchtimes, even for half an hour, this would help to break the monotony of his long, lonely day. If this is impossible, make sure he always has items on which to occupy his beak, such as pieces of apple or willow branches, and a rope swing for gnawing. Rotate his toys on a regular basis and leave the radio playing music in his absence.

M

Mimicry

All the parrot species with which we are most familiar as companion birds are essentially flock species, continually in vocal contact with other members of the flock. This is why, in a human household, they respond to human speech, and to other sounds. As pointed out by Sam Foster and Jane Hollander:

> parrots often imitate noises they hear in their daily interactions with us because they interpret them as contact calls. African grays are notorious for their microwave beeps and telephone ringing noises. Why those particular sounds? Perhaps because our grays see us answering the phone or microwave by physically going to those appliances. To us, we're taking dinner out of the oven or answering the phone, but to our grays we may be responding to the microwave or telephone's contact call. If it works for the microwave, it might just work as a contact call for the gray to bring us to him. (Foster and Hallander, 1999.)

Many parrots are able to mimic sounds, as well as speech. They can use this form of mimicry in an appropriate manner. A friend in California has a Timneh Grey Parrot called Alex—and a dog, a miniature Schnauzer. When the Timneh scratches himself, he often mimics the sound made by the dog when it scratches. When he wants more food—when the container in his cage is empty—he will make a crunching sound. He will also do this if he wants something Gwen is eating. He knows the value of hinting, here! When he is given fresh water or when he wants something that Gwen is drinking, he makes a gulping, drinking sound. 'Hey! Come here! Watcha doing?' is reserved for when he is in his cage and wants to be let out. Most amusingly, he says: 'No!' when he is about to do something forbidden!

Gwen and her husband have a certain whistle combination that they use when calling the dog in from outside. Alex mimics this and also uses it when the dog is barking or when they go to the door to call the dog themselves. He has only once used it incorrectly when Gwen went to the door for another purpose. She has not been able to discover what cues him but believes that it might be something subtle in her body language. (*See also* Talking.)

One of the most fascinating forms of mimicry among parrots is much rarer than vocal mimicry: imitating actions. Probably the best description of this relates to a Slender-billed Cockatoo (called Eastern Long-billed Corella in Australia). In the first volume of W.T. Green's three-volume classic *Parrots in Captivity* (published in 1884), he relates how one particular cockatoo was a great mimic of actions:

> daily he goes through the performance of pouring out tea. The sugar is first put into the cups—the action of putting it in, you understand—and then the tea is poured out; his beak being the spout of the tea-pot, and he makes the noise of pouring exactly, while pretending to do so ... One morning, from an old ladder that is devoted to his use in the garden, he watched the gardener clipping the laurels, and when he came in, we had the whole performance of clipping laurels gone through exactly, giving his head a little jerk with each snap of the shears—his beak was of course supposed to be those implements—and the sound was exact.

This form of mimicry tells us a lot about the powers of observation and memory retention of this cockatoo, to say nothing of its ability to act out what it had seen. The Western Slender-billed Cockatoo (this was the closely related form) is among the most intelligent and playful of all parrots, in my opinion, on a par with the Kea, New Zealand's mountain parrot. Behaviourally they have much in common.

Moodiness

Some parrots are undeniably moody—and this moodiness seems to be reflected in the time of day. Whether there could be some external influence would need to be investigated. A hand-reared Senegal Parrot named Charlie belonged to a lady who described him as 'the most comical, loving, sweet-natured little bird, full of joy and the most laughable little tricks, lying on his back and juggling his toys in his feet'—90 per cent of the time. But first thing in the morning and for a period in the afternoon he became totally unpredictable, with 'the most ferocious growl'. In the evening he was tucked under her chin, emitting soft little wolf whistles and making clicking and kissing sounds, having his head and neck tickled. He would stretch his neck as far as it would go, gazing up into her face and saying 'Give us a kiss'.

Moulting

The moult is a natural process whereby the feathers are renewed, usually on an annual basis. It varies in duration according to the species. Some have a fairly rapid moult (except for the wing feathers) which lasts a few weeks. In others, the process extends over a longer period. Wing feathers, in particular, are renewed over a period of months. In some parrot species, some wing feathers may be replaced at an interval of one to two years. The average period of the wing moult in the Galah, for example, is 160 days. Migratory species (those which move long distances in search of food) must be able to fly strongly at all times, so the moult is gradual.

The first moult in parrots occurs at between about five and ten months of age; smaller species at an early age, the large macaws at about eight to ten months. This can be a difficult time for the larger parrots, especially the large macaws. The stiff new feathers, still in their little sheaths, feel uncomfortable and might result in the parrot being bad tempered. A daily shower will help to loosen the shafts. Great care should be taken in caressing the parrot's head as this might be very tender. In this case no head scratching should be carried out. If a macaw's long tail feathers remain in their sheaths for a long time, spray them continually to soften them and, if necessary, gently remove the sheaths. In a dry atmosphere the sheaths may become so hard that they do not break away when the bird is preening, as would normally be the case. But take care! Most parrots intensely dislike the sensation of having the tail touched.

Providing a good diet is the best way to prevent moulting problems. Protein is required for the growth of new feathers; cooked chicken or other meat can be offered to a moulting parrot, or cooked pulses, such as beans and lentils, which are also high in protein.

Question
Why has my newly acquired adult Amazon parrot moulted out with dark edges to some of his feathers?

Answer
This indicates a dietary deficiency. A balanced diet, including vitamins and minerals, is essential for the production of healthy feathers. The diet must be very poor indeed to produce these symptoms. As you have not had the bird long, the condition probably reflects neglect and/or insufficient food before your Amazon came into your care. It is essential that your Amazon eats plenty of fresh fruit and vegetables as well as some cooked chicken. Most Amazons relish chicken.

Music

Most parrots are very responsive to music and to recorded bird song. They join in—chattering, yelling or even singing, according to the species. There is no doubt that they find certain music intensely stimulating. It is certainly a good way of relieving the boredom of parrots which must be left alone for long periods, whether or not they have a companion. Leave the radio switched to a music station and their day will be so much happier. The kind of music which is preferred varies but the response to different kinds will leave one in no doubt.

Amazon Parrots, especially the subspecies of *ochrocephala*, love to sing-along. The ability of some birds, notably Yellow-naped and Double Yellow-headed Amazons, is nothing short of extraordinary. A Yellow-naped which had belonged to a lady opera singer even made a record of 'I left my heart in San Francisco'. It was a near-perfect imitation of its owner's high quavering voice—and it never failed to delight me. Amazons, more than any other parrots, seem to be stimulated by music and especially by soprano singers. So don't be surprised if your Amazon becomes very excited at the sound of a high female voice—spoken or singing.

One parrot's reaction to recorded bird song was very interesting. The owner of a six-year-old Grey Parrot left a CD of bird songs playing whenever she left the house for a few hours, hoping that this would entertain her. One morning when the parrot had the freedom of the lounge she flew to the unit containing the radio/CD player and examined it closely. She lifted up the lid of the CD player, but stepped back as the lid gently returned to the closed position. Then, on tiptoe and with neck stretched, she lifted the lid as high as it would go, swiftly picked up the CD of birdsong and flung it on the floor! She then flew back to her cage. This might be interpreted to mean that she hated the CD. Perhaps she did! On the other hand, parrots in general and Greys in particular, love picking up items from a table, for example, and throwing them over the edge. They apparently enjoy watching things fall to the floor!

Mutilation, self

This is a very serious problem. It refers not to feather plucking but to mutilating the flesh, usually on the breast or on the feet. The parrot may actually bite a large hole in its flesh. This behaviour is most likely to occur with highly strung species such as the white cockatoos and Hawk-headed Parrots. The usual procedure is to take the bird to a vet who will put an 'Elizabethan collar' on it. This is a plastic collar which prevents the bird reaching any part of its body to mutilate itself. The theory is that when the wound has healed the collar can be taken off. However, this is only treating the wound and not the cause of the problem. On removal of the collar the problem may start all over again. While the collar is in place, it would be advisable to use a homoeopathic remedy such as valerian which has a calming effect.

The use of a plastic collar is extremely stressful and, in a parrot of exceedingly nervous temperament, could do more harm than good. I had a male Hawk-headed Parrot which mutilated its flesh on two occasions, once on the back and once by the crop. The latter area was naked from feather plucking over a long period. The mutilation seemed to be triggered by breeding condition. Nothing in the bird's environment had changed. I knew instinctively that the stress of taking this bird to a vet would have killed it and that he would not tolerate wearing a collar. I did not want to add to his stress by changing his environment in any way. Fortunately, on both occasions he stopped biting his flesh after a couple of weeks and the wounds healed. However, in other parrots the habit has been known to continue and sadly death has resulted.

N

Nervousness

Parrots which have not been hand-reared usually become nervous if closely confined. The current trend towards keeping noncompanion parrots in smaller and smaller aviaries exacerbates this problem. Many parrots, especially those taken from the wild, exhibit nervous aggression in small aviaries. In larger enclosures this behaviour disappears and the birds appear much more relaxed.

It is best to avoid eye contact with nervous parrots. Keepers of aviary birds should give some thought to how they move around their aviaries. Most flights have a service passage at the back from which feeding is carried out. Feeding should commence at the far end so that it is not necessary to pass a bird which has just been fed. This is because if you pass a nervous parrot when it is holding an item of food in its foot, it will drop it. Some people inadvertently make their birds feel very nervous by walking about with a catching net in their hand. Always hide it. Even calm birds hate nets. Many also hate gloves. Some will attack the owner's gloved hand although the bare hand is always accepted. Brightly coloured gloves are most likely to provoke an attack.

The tamest aviary birds will react nervously when a hawk, a heron or a species with a threatening silhouette flies overhead. Helicopters also come into this category. Moving hose pipes can cause a nervous reaction, presumably because of their perceived resemblance to a snake.

What can the owner of a very nervous companion parrot do to make it calmer? He or she could seek the help of a homoeopathic vet, or use Bach remedies (*see* Bach Remedies, page 86). Secondly, give the parrot a much larger cage.

O

Obesity

Just as in humans, companion cats and dogs, obesity in parrots is a major problem that results in an early death. In parrots the main causes are a poor diet, with an excess of high-fat foods such as sunflower seed and peanuts, and lack of exercise. Wing-clipped parrots are especially at risk. The owners of such birds must change the diet and the standard of care, seeking veterinary advice if necessary.

Don't know why I'm so overweight?
I only eat sunflower seeds!

Old Age

The ages to which parrots can live have been greatly exaggerated. Claims of 80 to 100 years often arise from a genuine misinterpretation. A bird which has been in the family over decades might be two birds, but a family member may not have been aware that one died and was replaced. He or she, perhaps as a child, can only remember the continued existence of a parrot. By the time most parrots reach their fifties (in fact, few do), they are in poor condition, with arthritis and, especially in the case of macaws, also with cataracts. Arthritis, as in humans, can be seriously debilitating. Even although an old parrot's feet may look normal, they cannot grip as efficiently as previously. This results in some parrots falling off the perch at night. To reduce the impact of such falls, the cage should contain only low perches. In severe cases, the parrot could be moved at night to a pet carrier with one very low perch. Arthritis in the wings will render a parrot unable to fly. Small parrots, including Cockatiels, can live up to 30 years and often appear to be in fairly good condition.

What can be done about arthritis? Little can be done to prevent it except to ensure the parrot is on a healthy diet that does not result in obesity. Perches of varying width in the cage or aviary help to exercise the muscles of the feet—and exercising the affected parts is beneficial. For an aged parrot that finds gripping difficult, six-sided wooden perches are helpful. Regarding actual treatments under veterinary care, non-steroidal anti-inflammatory drugs can help to relieve the pain. Some owners recommend the use of glucosamine/chondroitin/msm in liquid form, added to the food, for arthritic birds. However, caution is recommended as the results of long-term use are unknown.

Question
Could it really be true that Winston Churchill's Blue and Yellow Macaw "Charlie" was still alive at 104 years old? This story appeared in several newspapers in December 2003.

Answer
No! I unexpectedly came across the evidence that "Charlie" was not 104! One newspaper devoted a whole page to the story—

but it was not her photo that caught my eye. Here at last was a photo of Winston Churchill with his bird on his shoulder. *It was a Scarlet Macaw!* Even in a black and white photograph the difference is obvious. The upper mandible is light coloured, the breast feathers are dark and there are no prominent feathered lines around the eyes. The verdict was clear. Charlie was an imposter!

P

Phobic behaviour

'Phobic', in relation to parrots, is a term which is often used in American parrot circles. It refers to parrots which have suddenly become terrified of something—often their human companion. The owners usually say that they do not know the reason for this. One possible reason is a calcium deficiency in the diet. This can apparently affect the brain, resulting in an increased level of fear and/or aggression. Whatever the reason, the owner's attitude to his or her parrot will almost certainly change subconsciously, probably for the worse. This will make the task of regaining the parrot's trust even more difficult. No matter how tense and upset the owner might be feeling at the way the parrot is behaving, he or she must exude an air of calm. When a parrot is feeling extreme fear, if it feels that fear is returned—as well it might be if the bird has suddenly started to bite—trust-building is impossible.

One lady who wrote to me was honest enough to admit that her parrot had become phobic because she had had its wings clipped. She wrote:

Question
Wing-clipping has turned my lovable bundle of feathers and mischief into a nervous wreck. He is not feather plucking, but he has started nail biting. Even to get him to come to me is a struggle. Every time I look at him I feel so guilty. Will he ever trust me again?'

What are you in for?

Answer

This is difficult to predict. He might transfer his affections to another family member. If you were present when his wings were clipped he might associate this act with you for a long time. It is not advisable for the owner to be present when this traumatic event occurs. You will need a lot of patience and a very caring attitude. You have to win his trust all over again so your demeanour around him must be submissive. Avoid eye contact and move around slowly and quietly when you are near him. I would suggest that you spend periods sitting near him, perhaps watching TV or reading. Be very quiet and relaxed. Everything about you should be non-threatening. Above all, have patience and do not expect a rapid change in his behaviour. It could take weeks, or even months, before he trusts you

I'm out on good behaviour

again. You said to get him to come to you is a 'struggle'. Don't push him. He must want to come to you. He has to come to you on his own terms—not on yours.

Preening

Maintenance of plumage is of vital importance to flying birds, thus they spend a significant portion of each day preening their feathers. A parrot preens every area which it can reach. It will often invite its mate or sibling or a flock member to preen its head. This it does by lowering the head. This is also a submissive gesture. Mutual preening (also called allo-preening) has another function when carried out by male and female. It helps to strengthen the pair bond.

The extent of mutual preening varies according to the

species. It is indulged in for long periods by white cockatoos and to a slightly lesser degree by lories, macaws, conures, Amazons and other neotropical parrots—in other words, those in which the pair bond is strong. In other species, such as Eclectus, and Ringnecks and other *Psittacula* parrakeets, mutual preening does not occur frequently or it may never occur, depending on the compatibility of male and female. This has implications for the pet keeper because species in which the pair bond is not usually strong seldom tolerate having the head scratched. In contrast, cockatoos revel in it. In fact, they enjoy having all parts of the body caressed. They are the most sensual of birds.

A parrot which is closely bonded to its owner will preen his or her hair and face. My Amazon would preen around my eyes. I considered this to be a great compliment because preening close to the eyes denotes trust on the part of the bird being preened. Likewise, I would gently 'preen' the little feathers close to my Amazon's eyes.

Question
Why does my lorikeet rub her beak at the base of her tail when she is preening her feathers?

Answer
An oil gland is situated on the rump, near the base of the tail, in most parrot species. Some neotropical parrots, such as Amazons, *Pionus* parrots and *Brotogeris* parrakeets, lack the oil gland. It is also called the uropygial gland or preen gland. Its purpose is to produce a kind of wax which is rubbed on the feathers to help to keep them in good condition. The secretion is said to contain antibacterial and antifungal properties. Parrots have small preen glands compared with those of many species, such as waterfowl.

Predictability

This must apply to *you* in the way you interact with your parrot. Your behaviour must always be kind and consistent. If your parrot can rely absolutely on your reactions, its own behaviour will be much more predictable. (*See* also Moodiness.)

Prevention

The prevention of behavioural problems is so much easier than trying to cure them! Most such problems stem from lack of discipline and lack of training. Everyone understands that a puppy has to be trained, but it seems that there is not yet a general acceptance that the same is true of the companion parrot. (*See* Training.)

Punishment

Elsewhere in this book it is mentioned that trying to put yourself in your parrot's place, or trying to think like a parrot, will help you to understand it and its motives. If you can learn to do that, you will realise that punishment is at worst pointless and, at best, less effective than reinforcement in changing behaviour. Punishment is something that does not exist in the animal world; it is merely a human concept and not one which we should expect a bird or animal to understand. Instead of punishing (by whatever means) a parrot for being noisy, for example, we should be concentrating on talking to it and praising it when it is behaving well, occupied with its toys, for example, and ignoring it when it is noisy.

Tame parrots crave human attention. If you walk out of the room every time the parrot starts to scream, it will eventually come to realise that screaming has the opposite effect to that intended—which was seeking attention. However, this treatment is unlikely to be successful with cockatoos which have been force-weaned. In this case, this behaviour stems from a deep-seated insecurity which commenced at a very early age. It is a psychological problem of such seriousness that the cockatoo might be described as mentally disturbed.

Even if parrots could understand the concept of punishment, how would they know which act they were being punished for? We cannot tell them. Birds are creatures of such quick actions that by the time our 'punishment' has been put into force they are probably completing another form of behaviour.

R

Regurgitation

Courtship feeding occurs in most parrot species, based on the need for the male to feed the female while she is incubating eggs, because she does not have time to forage for food. It does not normally occur in the white cockatoos because incubation is shared by male and female and the male incubates for much of the day. He has no time to feed the female who can spend much of the day foraging for herself. In parrots the male regurgitates partly digested food for the female, and later for the female to feed to the chicks. A parrot which has a close bond with its owner will feed him or her. I see this as a compliment which should be accepted. Regurgitation is perhaps best known in Budgerigars, many of whom will feed their reflection in a mirror. Lacking a mate or a mirror, some male parrots develop strange habits, such as feeding their feet or under their wing.

The first time that regurgitation is observed, some owners become alarmed. One Amazon owner wrote to me describing how he had rushed his bird to the vet: he thought it was vomiting. So how can one distinguish between this natural act of a bird in breeding condition and a sick one? Regurgitation is usually accompanied by exaggerated movements of the head, either pumping or swirling, depending upon the species. This may or may not happen when a bird vomits but, when it does so, it has some obvious signs of ill health. Its eyes will look dull, the feathers will not be held sleek to the body, it will be quieter than normal and may show little interest in food. In contrast a bird which is regurgitating is usually very fit and bright-eyed.

Question
I have a hand-reared Yellow-crowned (Yellow-fronted) Amazon who is 18 months old. He is tame and healthy but almost every day he regurgitates about a teaspoonful of undigested food. It does not have a sour smell. He usually does this when he is playing with his toys or just amusing himself.

Answer
As your Amazon is healthy and this behaviour usually occurs when he is playing, it would appear to be normal—although 18 months old is quite an early age for regurgitation to occur. Some parrots do try to feed a favourite toy, rather than their owner.

Rings (Bands)

Closed rings (called bands in the USA) are used to prove that a parrot is captive-bred and to identify the individual bird. The ring usually carries the year of hatching and initials which can identify the breeder. Most parrots ignore the ring, which is put on at about the time the chick's eyes opened. If the correct size ring is not fitted it can be a source of danger, as the bird could become trapped by wire or a twig which enters the space between leg and ring. If the ring is too small, it is even more dangerous, as it could cut off the blood flow to the leg.

Another danger arises from breeders who use a ring which is not strong enough for the species. I witnessed an unfortunate example of this. The breeder had put a ring of a strength suitable for an Amazon parrot on the leg of a Green-winged Macaw. The macaw had crushed the ring into its leg, so that part of the ring was embedded. An emergency trip to a veterinarian followed. The ring was removed but the leg was painful for some days afterwards. Only stainless steel rings should be used for parrots with powerful beaks, especially cockatoos and macaws.

The types of rings used on parrots are as follows:

1. Split rings (identification only); plastic for very small birds, or metal. They might be used by a breeder who wanted to distinguish birds that had not been closed-ringed.
2. Sexing rings: less often in use these days. They were usually fitted by a veterinarian following surgical sexing. Females

were identified by a gold ring on the left leg and males with a black ring on the right leg. These rings are split and closed by a small pin.

3. Closed rings: as mentioned above, the correct size ring can be fitted only at about the time the chick's eyes open. If the ring is much too large for the species, it could have been fitted fraudulently at a later date.

Micro-chips are widely used these days to identify parrots. Either identifiable closed rings or microchips are essential for recording data on species listed on Appendix I of CITES (the Convention on International Trade in Endangered Species), as this information from breeders is required by most governments. In the UK, the Article 10 licence necessary to sell Appendix I species can be issued only to birds identified by one of these two methods.

Microchips are often the only means of identifying a lost or stolen bird without dispute, which is why many veterinarians recommend their use.

Question
Why does my Triton Cockatoo continually bite at his ring, to the degree that it is no longer round? Should it be removed—and how?

Answer
Some parrots, usually those with the most powerful beaks, get into the habit of playing with, then biting at the ring. There is a real danger that the ring could be crushed into the leg, causing a serious injury. This is especially likely to happen with the large macaws and with Grey-headed Parrots (*Poicephalus fuscicollis*). I would advise you to contact an avian vet as soon as possible. He or she will be able to remove the ring using a special tool. Do not try to do it yourself. Unless you know the technique, this could result in injuring the leg. Every owner of a ringed parrot should keep an eye on the condition of the ring. Parrots have lost a leg when a foreign object has become embedded between leg and ring, causing the leg to swell up. This is very rare but the danger exists.

If it is necessary to have the ring removed, keep the details of the ring number (if they are still legible). For security purposes, the parrot should be microchipped when the ring is removed. Note, however, that a ring is unlikely to help in the identification of your parrot if it is stolen, because most thieves cut off the ring. They would be unable to locate the microchip, so this method of marking is much safer. It is recommended instead of ringing for the large macaws and for cockatoos and Grey-headed Parrots.

S

Screaming

Excessive noise is a problem which most parrot owners—
certainly those of the large species—will experience at some
time. In many cases the problem is so serious that the unfortunate parrot will change hands many times and live an unhappy
life. In other instances, the problem is a temporary one.
Screaming, like feather plucking, is an indication that there is
a fundamental problem. This should be addressed by considering all aspects of the parrot's care. These include
occupational therapy (keeping him busy!), a stimulating environment (perhaps where he can look out of a window), a diet
with a variety of colours and textures and tastes, and plenty of
time out of the cage.

It is extremely important that people who live with parrots
understand the various factors which cause this problem. They
can then try to eliminate these to discover the reason.

1. Attention-seeking

This behaviour is most common in parrots which have been
hand-reared and expect a high level of human attention.
Cockatoos are infamous for this: because their voices are so
loud, it can mean that they are impossible to live with, unless
the situation is corrected. The therapy is two-part: training (*see*
Training) and providing a more stimulating environment. A
parrot which is busy with toys and items to bite and destroy, and
which is not kept permanently in the same position, is much
less likely to scream from sheer boredom.

One reason why parrots become screamers is that their
owners inadvertently reward them for screaming. Parrots soon

realise that they gain immediate attention with this tactic, even if the attention is short-lived and results in being covered up. Look at it from the parrot's point of view. When it is quiet in its cage, playing, feeding or resting, it receives no attention. But as soon as it screams, humans come running. Right from the start, the new parrot owner must acquire the habit of talking to his bird whenever he passes the cage—except when the parrot is screaming. One reason why parrots scream is because they think they do not receive enough attention. Talking to your parrot frequently, and perhaps rubbing his head as you pass, helps to prevent the start of the screaming habit.

Parrots which are left alone all day may, not surprisingly, become very demanding. The fact is that leaving a parrot alone for hours results in a poor quality of life. The parrot rebels at this by screaming for attention. While it is true that a person's circumstances may change, making this unavoidable, in my opinion, no one should buy a single large parrot knowing that it will be left alone for hours on end for five days a week.

One lady reported that her 20-week-old Grey Parrot had been trying her patience with his continuous squawking when she took him out and he was sitting on her shoulder. She had been working on night shift for two weeks. At 18 weeks old the Grey must have felt as though he had been abandoned by his favourite person (only her daughter was at home in her absence). Young parrots need a secure, unchanging environment in the post-weaning stage. The continuous squawking was probably due to anxiety—fear of being left alone. A person who understood the sensitive nature of Greys would not have bought a very young one knowing that soon her circumstances would change. This is a case of too little thought and understanding being given to the purchase of such a vulnerable creature. Would someone buy a puppy, knowing that soon after it would be left alone for hours?

Question
I have an increasing problem with my 13-year-old Blue-fronted Amazon. My father bought him for me five years ago. I am fully aware that Amazons have a reputation for being noisy but Jasper has now gone beyond a joke. And he is getting worse. I tried

putting heather oil in his water, as suggested in a bird magazine, but to no avail. I am now desperate for help before complaints from family and friends force me to sell him. This would not be a good idea as he is a one-person bird who will tolerate no one but me. I have tried to teach him by covering his cage when he screams but this has only resulted in him being quiet when he is covered. He starts screaming again when the blanket is removed. He is not short of attention. He even screams when he is with me. I have tried to teach him that if he screams when he is out he will be put back in—but this has not worked either.

Answer
This was difficult to provide with the given information. I telephoned the owner and obtained more information. I discovered that the Amazon had previously been kept in a cage in a caravan and never let out. Perhaps he got into the habit of screaming for attention. Currently he is in the house with her father during the day and usually screams in the morning. He sometimes attacks her mother and her father. When she takes him upstairs to her bedroom he is quiet. This suggested that one reason for his screams is jealousy. He is jealous of people who give attention to his young owner. I therefore asked: 'Does he try to feed you?' She said he did. This indicated that he regarded her as his mate and is jealous of people who come close to her. She should therefore avoid these confrontations.

The second step to take was to keep him busy in her absence. I suggested branches containing hawthorn berries as well as toys. The twigs provided kept him amused for a while but they were too quickly demolished. It was the various toys which kept him quiet. He had hours of amusement out of them and, as a result, the noisy behaviour has diminished.

It should be noted that Amazon parrots, especially males, can be very noisy. Mature birds whose history is unknown may never have been trained in previous homes. It takes time and patience to change the habits of many years. In a case where this proved to be impossible, one solution for a parent-reared bird might be to build an aviary and, if the temperament of the male was suitable, obtain a female of the same species. Unpaired males are at their noisiest during what would be their breeding season.

Question

Why does my Amazon become very noisy and demanding whenever I invite friends round for dinner? I live alone with my Amazon who is a very good, well-behaved bird most of the time. I always give her something from my plate so the reason is not that she wants something that I am eating.

Answer

The key is in the fact that you live alone. Your Amazon is used to having your undivided attention and she is jealous of the time you give to your friends when you invite them to dinner. Distraction techniques such as giving her a fresh-cut willow branch or her favourite food will only be successful for a short period or, perhaps, not at all in the circumstances. What she really wants to do is to join the party: to come out and sit on the table. However, unless your friends are parrot lovers they will probably stop accepting your invitations if you do this!

2. Fear and Stress

An unknown or threatening item introduced into the vicinity of a parrot's cage can make it fearful and noisy. This reason may be overlooked because to human eyes the item is inoffensive. The reader of a pet column in a Sunday newspaper asked for advice on a nine-year-old Cockatiel which had recently become noisy and aggressive. She had 'put a piece of string in his cage as a pacifier' and wanted to know if there was anything else she could do to stop the screaming.

The published answer from a vet suggested that the Cockatiel was probably 'bored out of his mind' and that a piece of string was no substitute for a mate. The advice was to give him space to fly, plenty of human company and to consider introducing another Cockatiel.

The first question I asked myself was why a nine-year-old Cockatiel's behaviour had suddenly changed. After all, it had been sexually mature for eight years. If the problem was lack of a mate, it is unlikely to have taken eight years for this to surface. I would have asked the same question as related in the example below. A vet should also have been concerned about the bird's health. A behaviour change could be caused by pain. The bird needed to be examined by an avian veterinarian.

3. Noisy environment

Many people inadvertently encourage their parrots to be noisy. A radio or television blaring out music is an invitation to join in. Turn it off and the parrot may instantly be quiet. Simultaneously dimming the lights will also encourage silence.

4. Attempted communication with carer

Owners should learn to interpret the kind of raucousness caused by a parrot that wants to be let out or is asking for something to eat and to distinguish it from a parrot which is calling to its owner due to insecurity. A young or recently acquired parrot, or one which has been poorly socialised and is just starting to bond with its owner, will be feeling great anxiety. It will naturally call to its favourite human companion as soon as he or she moves out of sight. This is the contact call which is such an important part of life for a sociable bird like a parrot. If its contact calls are not answered it will keep calling—or yelling. This places the owner in a dilemma. If he or she returns to the room to comfort or placate the bird—or to tell it to be quiet—the parrot soon learns that calling is an effective way of bringing back the person. The owner has to distinguish between the manipulative parrot and the one which is feeling so insecure it genuinely needs the vocal contact. In the latter case, calling its name from the next room may help to calm it. Alternatively, the owner might use a special whistle as a 'contact call'.

Question
Why does my parrot become so noisy during our meal times? He is kept in the living room, where we eat.

Answer
Feeding is a social activity for parrots. I would suggest that you replenish his food dish (you can give smaller amounts twice a day), or give him fruit or other fresh foods when you sit down to eat. Many parrots instinctively go down to the food pot when they see their human companions eating. If that pot is empty, they will scream to be fed. Or they might be yelling to tell you they would prefer what is on your plate to the food in their dish.

Dawn Chorus

5. Communication with other birds

This is natural behaviour which is unlikely to be modified. In the wild parrots keep in touch with each other over quite wide distances by calling. Parrot keepers with birds inside and outside the house can thus expect some calling between similar or related species. My tame Black-capped Lory (*Lorius lory*) was at his noisiest when returning the calls of the lories in my outdoor aviaries. Frankly, the best thing to do at this time was to leave the room, because as long as another bird was calling to him he was unlikely to be quiet.

6. Contact call for out-of-sight human

Lories and lorikeets are nectar-feeding parrots. Most of them are flock species but pairs remain in close contact within the flock.

Question

My husband and I bought a Green-naped Lorikeet (*Trichoglossus h.haematodus*) nine months ago. She is very affectionate and bright and spends a lot of time out of her cage. She rules the dogs with a rod of iron. The problem is that she always wants to be with my husband, who works at night. She calls all the time until she wakes him up. What can I do to stop this?

Answer

Your lorikeet considers herself to be paired to your husband. Lorikeets are among the parrots which form very strong pair bonds. It is natural for the members of a pair to call to keep in contact when they are out of sight of each other. Most of the time they do everything together. I believe that it would be very difficult to alter this instinctive behaviour which, in the circumstances, has become extremely annoying. There are two possible solutions. One is that your husband takes the lory up to bed with him. There could be a small cage in the bedroom. She might be satisfied with a few minutes with him before he goes to sleep, if she can then see him.

If this solution is unacceptable, I would suggest that you look for a male Green-naped Lorikeet and place them together in an indoor or outdoor aviary with a nest-box. When she bonds with the lorikeet, her noisy behaviour will subside.

SOLUTIONS

What solutions are there for dealing with noisy parrots? The usual one is to cover the cage. This should be as a last resort only—perhaps when you are speaking on the telephone and simply cannot hear the caller. Many parrots call out at this time because they resent the attention you are giving to the telephone.

At other times ask yourself why your parrot is being so noisy before you take action. Often he is trying to tell you something and you are not listening. If he is just being noisy for the hell of it, try the distraction technique—something to keep him busy. However, this must vary and not occur on a regular basis, otherwise it will seem like a reward.

Writing in *Parrots* magazine (October/November 1998) Julie Brinklow described a very effective method of stopping her cockatoo screaming. When he started, she put in earplugs, making sure he saw what she was doing. She was then able to ignore him totally—no eye contact, nothing. He soon became aware that the use of earplugs meant that he would be ignored, thus there was no point in screaming. The earplugs are seldom used these days. When screaming did not work, he tried other methods of attracting attention, such as throwing down food dishes, banging his toys against the cage bars or disappearing under the paper in the cage—but they were equally ineffective. His owner wisely advised: 'Just ignore them when they scream and reward them when they are being good.'

Eb Cravens from Hawaii has a couple of interesting suggestions. He teaches the young parrots he hand-rears to sleep in a cardboard box at night. If they are very noisy during the day he puts them into the box and stuffs a T-shirt in the entrance for ten minutes. When the cloth is removed, they have calmed down. Such a sleeping box for small parrots causes them to sleep for an extra 30 to 60 minutes in the morning. This would work well with species which love to go into holes and boxes, such as conures, caiques, lories and lorikeets. It would be disastrous with Greys and other species which do not naturally seek out holes. Large parrots have their cage covered to 'delay sunrise' and keep them quiet early in the morning.

A solution which certainly works with some Amazons is to teach them to sing. Eb Cravens suggests: 'Merely join your parrot's screaming with operatic "la-la'las" for a few months and watch the results.' Many parrots greatly enjoy the interaction with a human who is singing to them. Some species, especially Greys, readily learn to whistle and are very responsive to whistles. This might be another solution to divert the attention of a noisy bird.

Parrots should never be punished for being noisy (*see* Punishment). It is such a fundamental part of being a parrot— and punishment is seldom or never effective in correcting any form of undesirable behaviour.

Sexual behaviour

The antics of parrots which come into breeding condition can change to a degree that can become very alarming to the owner. If he or she is not familiar with the behaviour which is associated with sexual maturity and the desire to breed, it is likely to go unrecognised. It is not infrequently mistaken for ill health. One owner of a two-year-old Amazon parrot was very disturbed to see the bird walking around the top of its cage every evening, fluttering its wings and making squeaking noises. It became very 'flustered'—and nothing would distract it. He thought it was having a fit. This is typical behaviour of an Amazon in breeding condition—probably a male, in this case. Expect this to happen in late spring.

A common assumption of bird owners when they observe this milestone in their pet's life is that the bird needs a mate of its own species. If it is contented, and bonded to its owner, nothing is further from the truth in the majority of cases. This applies especially to hand-reared birds which have never known the companionship of their own species. Many simply do not know how to behave and may react with aggression to another bird. The emotion most likely to be apparent is extreme jealousy (*see* Jealousy). A potential mate can cause many more problems than it solves.

Parent-reared and wild-caught birds are different, as are species which seldom form a strong bond with a person, such as lovebirds. Certain smaller parrots of a highly sociable nature, such as lorikeets and conures, might readily accept a mate, regardless of whether they are parent-reared or hand-reared. Hand-reared males of certain species—white cockatoos and *Lorius* lories, such as Black-caps—potentially pose a lethal threat to a female. Their reaction may be to attack with such speed that the unfortunate female stands no chance.

Under normal circumstances sexual behaviour or egg-laying by pet birds is a phase which will pass. It is not so disruptive to the human/parrot relationship as is the introduction of a potential mate for the parrot. It should also be borne in mind that most pet birds have been kept in quite small cages and unless they have had a lot of wing exercise, a female is not in breeding condition. If she were overweight or unfit, egg-laying could lead to her death.

Case history
This almost happened with my own cherished Yellow-fronted Amazon, more than 20 years ago. I unwisely decided that she needed a mate. Every May she came into breeding condition, shivering her wings and making little crying sounds. She was placed in an aviary with a male who was very interested in her. That interest was not returned. But she was excited over the availability of a nest-box. She would sit outside, shivering her wings at me, as though inviting me to join her inside. It was the presence of the nest site which caused her to ovulate—not the presence of the male.

Early one Sunday morning she laid an egg. I saw with absolute horror that she had a prolapsed oviduct. This means certain death unless dealt with speedily. I was fortunate that my vet had a Sunday morning surgery. He attended to her quickly and made a purse-string suture which allowed her to defecate but also retained the prolapsed organ. Needless to say, I never again allowed her near a male or an aviary.

Ex-pets in a breeding situation
This near-disastrous incident taught me a lesson I have never forgotten. A pet bird is not breeding fit. Most heavy-bodied parrots such as Amazons which have spent years in a cage would need months of exercise in an aviary to tone them up for breeding. It would be a different matter for a more slender species. Even so, it should not be assumed that a caged Ringneck Parrakeet, for example, can be put outdoors in an aviary, given a mate and a nest-box and is ready to breed.

The worst scenario is when someone persuades a pet bird owner that his or her hand-reared bird would be better off in a breeding situation. Removed from the domestic environment to an aviary, the poor bird is totally bewildered. In this frightened and submissive state of mind, it is vulnerable to attack from the potential partner. Of course, some former pets do go on to make good breeding birds. The transition can sometimes be achieved, usually by a sensitive and caring person who is aware of the problems involved.

Question
Can you tell me what is wrong with my mother's Green-cheeked Conure? Lately it has been rubbing its vent on the perch.

Answer
There is nothing wrong with the conure. The behaviour you describe originates from a desire to breed. Without a mate it is trying to mate with its own tail or with the perch. Such behaviour is often seen in single birds at the onset of the breeding season. It will probably last a few weeks. Just ignore it and do not let the conure find that this could be a way of attracting attention to itself.

Question
Recently my Cockatiel, a believed female, has started behaving in a strange way. She stands with her head low and her back arched, at the same time making little whistling noises. This usually happens when she is with my boyfriend. I wondered if this was something to do with breeding behaviour—but she is only four months old.

Answer
The behaviour you describe is typical of a female soliciting copulation. Cockatiels mature early—nevertheless your bird is quite precocious. She obviously considers herself to be bonded to your boyfriend.

Question
Every year my Grey Parrot, aged about ten years, sex unknown, goes through a period of ripping up the newspaper in the bottom of his or her cage. This period lasts from November until about January. At the same time he or she becomes very amorous and tries to feed me. Is there anything I can do to stop this behaviour?

Answer
I am not sure why you would want to stop it. Many adult parrots come into breeding condition, despite the lack of a partner of their own species. In fact, your parrot regards you as his or her partner. The behaviour you describe is quite natural and is not causing any harm or stress to anyone. The shredded newspaper is a slight inconvenience—but surely a small price to pay for a cherished pet? If it really inconveniences you, try reducing the hours of daylight. Either cover the cage early in the evening or move your parrot into a quiet dark room for the evening. I would not really recommend this course of action because you are depriving your parrot of time with you. He or she could resent this, leading to feather plucking.

Sexuality—human and psittacine
A letter in a Sunday newspaper posed an interesting question: can parrots distinguish heterosexual and homosexual people? It went like this: 'My in-laws used to have a Yellow-naped Amazon Parrot which was very flirtatious towards women and homosexual men, but attacked heterosexual men and lesbians. Being homosexual, my partner and I used Lorre as a discreet device to discover whether some men were as heterosexual as they appeared to be. It was funny to see how the parrot responded to certain husbands. We think Lorre's instincts were almost perfect. The men he flirted with have all, but one, "come out" after divorce.'

To believe that a parrot would discriminate between gay and straight men is to anthropomorphise! I believe that the Yellow-naped Amazon was subconsciously responding to the fact that homosexual men are generally more sensitive than hetero-sexuals and their body language is less aggressive—often more on a par with that of a woman. They therefore usually approach birds in what the bird perceives as a less threatening manner. Of course heterosexual men do not intend to be threatening—but their demeanour is different.

Psittacine relationships can be just as complex as human ones. Sometimes two parrots of the same sex will form a strong bond, despite the presence of birds of the opposite sex. This often happens with young birds, before they are sexually mature, but equally it can happen with mature birds. On no account should a bird of the opposite sex be introduced into the same aviary as the other two. It is likely that it would be attacked as an intruder, just as would occur if the 'pair' were a male and a female. The bond between two birds of the same sex can be just as strong as that of an opposite-sex relationship and splitting them up would be just as stressful to the parrots involved. Their happiness should come first.

Question
I recently bought a Blue-fronted Amazon, a DNA-sexed female, who is nearly two years old and was previously a family pet. I have always wanted an Amazon but to my dismay she has taken to my husband and always wants to be with him when she is let out. I have now been told that parrots always prefer people of the opposite sex to themselves. Is this true?

Answer
If a survey were to be carried out, it might reveal that the favourite person of most parrots was of the opposite sex but I doubt that the indication would be overwhelmingly conclusive. Past experiences can influence a parrot's attachment to men or women. An example of this was a female Grey Parrot, a rescued bird who was badly treated by a previous owner and disliked women. She took an immediate liking to the man of the family and fed him by regurgitating food into his hand. Slowly she was accepting his wife to the point that she would take food from her—but would not allow herself to be touched. It might be that in your Amazon's previous home it was the man of the family who paid most attention to her. On the other hand, I believe that many parrots prefer women because their voices are more attractive, and often more soothing, than a man's voice.

Shoulders, sitting on

Because parrots prefer a high vantage point to a lower one, a tame one will automatically try to climb from your hand to your shoulder. This might be acceptable with a small species like a Budgerigar, but it poses problems in the case of the larger parrots. The first reason is that it makes you vulnerable to being bitten. You cannot see the danger signals that might precede an attack (such as the flashing eye) and some of the most tender

Just stop singing and give me the grape!

parts of your anatomy—cheeks, eyes and ears—are close to the beak. A usually reliable parrot that would not bite its owner under normal circumstances will bite if someone approaches too closely when the parrot is perched on a shoulder or on a hand. This is due to misdirected aggression; it cannot bite the intruder so it bites the nearest person, instead.

The second problem when a large parrot is perched on your shoulder is that its eyes are above your eye level. This gives it a psychological advantage; it feels superior to you because of its elevated height. For this reason it might be difficult to control and refuse to step up'. Instead of stepping on to your hand, it will move towards the middle of your back, where it will be very difficult for you to retrieve. If you ask someone else to remove the bird, there is a good chance that he or she—or you—will be bitten in the process.

Sleep

It is very important that companion parrots receive enough sleep. Most of them reside in the most lived-in-room in the house. In most households the television is not turned off until late in the evening. If the parrot is in the same room, the cage should be covered by about 9.30 pm and the TV volume should not be loud. Even though a parrot may be dozing during the day, it still needs good quality rest at night. This is extremely important for a young bird. This fact cannot be over-emphasised. Failure to ensure that it had sufficient rest was almost certainly responsible for the death of a young hand-reared Grey Parrot only a few weeks after it went to its new home. Perhaps it was too tired and disturbed by constant noise and activity to eat properly. In any case, this should have been noticed. It is very sad to think that some parrot owners are either too uncaring or too ignorant to notice. Lack of sleep can cause parrots to become nervous, irritable and even nippy.

Question
Why does my Grey Parrot often sleep with one eye open?

Answer
This may indicate caution or suspicion. Do you, for example, have a cat which could suddenly jump on to the top of the cage and frighten your bird? Sometimes birds which are just dozing during the day sleep with one eye open—a natural instinct which helps to keep them alert and aware of the presence of potential predators.

Researchers at Indiana State University studied the sleeping habits of birds, using Mallard. The ducks slept in a line while their brain patterns were filmed and monitored. The ducks in the centre of the groups tended to sleep with both eyes closed while those at either end were more likely to keep one eye open. Electrical readings revealed that even although the ducks at the end of the line were sleeping, they were still alert—as was the half of the brain controlling the watchful eye. The half of the brain controlling the closed eye was asleep.

This technique has been called unihemispheric slow-wave (USW) sleep. During this period, activity in the alert half of the brain increased when the birds were led to believe that they were under attack—by showing a predator on a video screen. They awoke within one tenth of a second and were in escape mode. This was believed to be the first evidence indicating that an animal can behaviourally control sleep and wakefulness simultaneously in different regions of the brain. It probably applies to most bird species. In order to rest both sides of the brain, the ducks would stand up every hour or so, turn round and sit down again so that the other eye was used to keep a watch.

I believe that this shows we are only on the threshold of discovering facts about birds which indicate that they are creatures far more complex than most of us could ever imagine.

Smell, sense of

Most parrots probably have little need for a well-developed sense of smell; after all, in the great outdoors most odours disperse fairly quickly and would be of little practical use, except possibly for lories that need to locate flowering trees. The apparatus for detecting smell is present in the nasal passages of all birds. Based on the relative size of the brain centre used to process information on odours, physiologists expect the sense of smell to be well developed in rails, cranes, grebes and nightjars and less developed in passerines, woodpeckers, pelicans, and parrots.

The results of what were described as the first behavioural investigation of olfactory capability in any parrot species were published in 2003 (Roper, 2003). Two Yellow-backed Lories (*Lorius garrulus flavopalliatus*) were able to distinguish painted grey tubes containing water from tubes containing nectar that were scented with colourless odour such as patchouli, in the form of essential oils. The author noted that in 1972 an attempt to record physiological responses to odours in a Yellow-fronted Amazon failed. He concluded that nectar-feeding parrots, such as lories, might use olfaction to evaluate individual flowers from close range, a situation mimicked in his experiment.

Question
Sometimes when I am preparing dinner in the next room to my macaw, he calls out in an excited way. Can he smell the foods I am preparing?

Answer
Take note of which foods you are preparing when he calls out. If he does so consistently when you are cooking chicken, for example, it would seem that he could distinguish this odour. If he calls out when you are cooking a wide range of foods, it might just be that he can hear the sounds that indicate food is being prepared, and calls in anticipation of what you might give him from your plate.

Smoking

'Passive smoking' is even more of a danger for parrots and other birds than it is for humans. Unlike mammals, birds have a respiratory system which enables them to absorb oxygen when they are breathing *out*, as well as when they are breathing in. This means that they are more efficient than mammals at absorbing toxins in the atmosphere. A bird does not rely on its lungs to breathe, as do mammals. Instead it uses a series of air sacs which take up much of the space within the body cavity which is not filled by the organs.

Inhaling cigarette smoke over a period of several years can literally prove lethal to parrots. This is no exaggeration. I know of a much loved parrot who would sit on his owner's shoulder every evening. Unfortunately, the owner was a heavy smoker. The parrot died from a respiratory condition (its air sacs were badly affected) when it was only five years old. This was a sad waste of a precious life.

Coughing and sneezing might be the first indication that cigarette smoke is having a harmful effect. On the other hand, there might be no outward indication until the parrot suddenly becomes ill. Sinus infections, rashes on the bare skin of feet and cheeks, and air sac infections are some of the serious results of exposure to cigarette smoke.

Smokers should also be aware that if a packet of cigarettes is left lying about, the packet will be chewed up if found by the parrot. If it also ingests tobacco the result could be vomiting, diarrhoea, fits and even death.

Sneezing

The odd sneeze is no cause for concern but a parrot which sneezes often and regularly needs veterinary attention. It could have an allergic reaction to house dust or to cigarette smoke, for example. It might be disease-related—psittacosis (chlamydiosis)—which could be accompanied by a discharge from the eyes or nostrils. A lack of Vitamin A in the diet often causes disease of the respiratory system (*see* Smoking).

Question
My African Grey Parrot often sneezes. Why? My sister says that her Grey never sneezes, so I am worried that there is something wrong with my bird.

Answer
Your concern may be well founded. If your parrot is fed mainly on a seed diet, there is a strong possibility that it is suffering from a Vitamin A deficiency: this is very common in this species and other large parrots. It results in a thickening of the sinus lining, making it prone to infection. A chronic sinus infection would need long and expensive treatment to cure, resulting in much discomfort to the bird. The nasal discharge from a sinus infection can result in enlarged nostrils and even an unsightly channel in the upper mandible.

Other diseases can also cause sneezing, thus an avian vet should be consulted immediately. Note that if smoking is permitted in your household on a reglar basis, the smoke can irritate the nostrils and cause sneezing. An allergy could have the same result (see pages 38-40).

Socialisation

In a companion parrot context, socialisation usually refers to young hand-reared parrots having good, trust-forming contact with people at an early age. However, it might be that contact with other parrots is equally important, whether a parrot is destined to become a pet or a breeding bird. If a young parrot is reared alone, whether by its parents or by hand, it can develop behavioural problems unless it has the companionship of is own species very soon after it is independent. In the company of other birds it might not be accepted. A colony of Spectacled Parrotlets (*Forpus conspicillatus*) was studied in a large outdoor aviary and the findings were published in a scientific journal (Garnetzke-Stollmann and Franck, 1998). Flock number varied between ten and 30 birds. The flock's activities tended to be synchronised. At any given moment most birds would be foraging, preening or resting. If something frightened the flock members, they would scatter, then remain silent. When the danger had passed they would make contact calls and reassemble in the perching tree. The newly fledged young were observed. It seemed that play was essential for social learning; sibling relationships were apparently crucial in this learning process. Young birds were often observed fleeing from other birds, even their parents, but seldom from a brother or sister.

They began to form their first pair bonds a few months after fledging, but these relationships did not end the close ties between siblings. Strong pair bonds were not usually formed until the parrotlets were about one year old. Only then did a bird cease interaction with its siblings. However, if the pair bond was broken, the sibling relationship could be re-established. The importance of sibling relationships could be seen in the fate of single fledglings. Lacking brothers and sisters, they attempted to maintain a close relationship with their parents. This was sometimes successful once the parents had passed through the phase of distancing themselves from their young. A few single youngsters even helped to rear younger siblings.

Six single youngsters were studied. Four had trouble forming or maintaining pair bonds, and one female remained socially isolated. Only one was successful in forming an exclusive relationship and reproducing. He was fortunate in being able to attach himself to two young females that had fledged about the same time as he did.

Clearly, lack of opportunity to socialise with other members of their own species at an early age can affect their breeding potential. In my experience, young males may become excessively aggressive or bullying and young females may be very difficult to pair up when mature. It can also profoundly affect their personality, and therefore their pet potential.

Case history

Two friends have two Grey Parrots which were bred from two different pairs. They are seldom left alone and they have long periods of full-winged freedom within the house. Both are males and their personalities are quite different. During adolescence, Gilbert, the Grey who was obtained first, was extremely temperamental for months. He objected to being put back in his cage and his owners thought he would never improve. But he did. He is now well behaved although he will nip occasionally if he is not pleased at what is going on. In contrast, Gus sailed through adolescence, with only a brief difficult period lasting a couple of weeks. Gus has the most wonderful temperament and can be handled by strangers. He never bites. If he is displeased he will put his beak around a finger and push the finger away. He is an exceptionally intelligent bird.

Why should the temperament and personalities of these two male Greys be so different? Both were reared by the same excellent breeder and both have always lived in the same household? We can never be sure of the reason. But it is very interesting to look at the early weeks of their lives. Gilbert was only nine weeks old when he went to his new home. He at once bonded with the man of the house, Ian, who works from home. When Gilbert was 12 weeks old Ian went away for two weeks. Gilbert was very upset and it was difficult to persuade him to feed. A few months later Gus joined the household. He bonded most strongly to Tracy. Tracy teaches, so she is absent for some hours on week days, but Gus has always had Ian and Gilbert's company. Gus had the perfect secure upbringing, including being reared with two siblings. Gilbert was a single chick. Gus had every advantage in life, especially early secure socialisation, and this may well be why his temperament has always been so stable and loving.

Stereotypic behaviour

Apparently meaningless behaviours that are repeated in a persistent and fixed manner are described as stereotypic. In caged animals pacing the perimeter of the cage or enclosure is typical. This can indicate some form of mental disturbance. For example, a Moluccan Cockatoo which was cruelly kept for two years in a 60cm cage in a dark coal bunker would repeatedly climb round and around in a similar small area when placed in a very large cage. Milder forms of repetitive, abnormal behaviours might be seen in parrots in small cages with nothing at all with which to amuse themselves. On a number of occasions I have seen parrots in small cages repeatedly bobbing up and down on the perch—but not in display. The response by onlookers has been: "Oh, look! He's dancing!" They have interpreted the parrot's actions in the most anthropomorphic way possible. In fact the bird is frustrated at being confined to a small space and this might be a thwarted desire to fly, endlessly repeated.

Sun

Many parrot owners have the mistaken idea that parrots love to sit in the sun. Most of them hate it. Aviary birds often tolerate early morning sun but they will disappear inside the shelter as soon as the sun gains strength. Remember that most parrots are birds of the forests and rarely encounter the sun. Parrots kept indoors should *never* have to tolerate direct sunlight. If they live near a window, there should be a blind over the window so that the sun can be blocked out when necessary. I recall on one occasion on a cloudy day, I left my Amazon on her stand in front of a window. I was working in another room and failed to notice when the sun came out. She could have been on her stand for half an hour or so in direct sunlight. When I realised her plight I rushed in to retrieve her. In my absence she had plucked out some of the feathers on her abdomen. This was the only occasion on which she ever plucked herself.

She disliked being outdoors in her cage for more than a few minutes. If the sun should reach the cage, she would bite on the bars in distress and make a short loud sound which was never heard at any other time. A friend has a Grey who hated going outdoors until she hit on the idea of putting her cage under a large sun umbrella. Whether this dislike of the outdoors is partly fear of the open sky is difficult to assess.

Not all parrots hate the sun. A few species are sun worshippers.

Top of the list must be the Vasa Parrots (*Coracopsis*) from Madagascar. They assume bizarre postures, with wings or wing outstretched, practically lying on one side, to take in as much sun as possible.

T

Tail-bobbing

Normal breathing is accompanied by hardly perceptible tail movement. If tail movement is marked, the lower respiratory tract may be diseased. Consult a vet immediately.

Talking

The key to teaching a parrot to 'talk', that is, to mimic human speech, lies in spending much time with it and talking to it frequently. A parrot which lives in a household where there is little conversation cannot be expected to be a good mimic of human speech. In addition, there must be a good relationship between bird and owner. The parrot should be confident and unafraid. An unhappy or nervous bird is unlikely to learn to talk.

Many people have unrealistic expectations regarding how long it takes to teach a parrot to repeat words. Most Greys, for example, do not learn to talk until they are about one year old. In contrast, some large macaws are saying their first words before they are even weaned. These are likely to be such expressions as 'Hello!' and 'Come on!' which are short and frequently heard. Greys and other parrots will start to mimic by muttering sounds which are unrecognisable as words. They are practising. This phase might last for some weeks before a word is clearly pronounced. On the other hand, some adult birds, especially Greys, learn new words and phrases within the space of a few days and do not need to practise the sounds.

A question which is often asked is whether male or female parrots make the best talkers. In certain species, including Budgerigars and Cockatiels, there is no doubt that males are the most talented talkers, and some females never learn to talk. In most species, however, talking ability is not related to the bird's gender. Neither does ability have any connection with size. Many large parrots, especially cockatoos, never learn more than a few words. Yet Budgerigars and the tiny parrotlets (*Forpus*) species can acquire a vocabulary which is amazing in its extent. Obviously, small species are unable to reproduce words with much volume and with virtually no accuracy of tone. Grey Parrots excel above all other species in the faultless accuracy of the spoken word, to the degree that it may be impossible to differentiate bird and owner. Certain Amazons, such as the Yellow-nape, come a close second. Cockatoos and macaws rarely have the ability to reproduce the human voice so faithfully.

Some people find it amusing to teach their parrots to swear. Just consider, however, that should you be unable to keep your parrot some time in the future, the fact that he swears may make him unsaleable and impossible to re-home. It caused one parrot to be 'sacked' in 1999. A Yellow-fronted Amazon called Percy was cast as Long John Silver's parrot in a pantomime production of *Treasure Island*. He learned his classic line 'Pieces of eight' and played his part well. However, during a dress rehearsal he turned the air blue when he suddenly came out with some language which was even more colourful than his plumage!

The mechanism of speech

In the whole of the avian kingdom, only certain parrots, mynahs, starlings and crows have the ability to reproduce the human voice with accuracy. One reason why parrots are good mimics is because they have a fleshy tongue. In many birds the tongue is very thin; in toucans, for example. A fleshy tongue is just one of the requisites for vocal mimicry. It is only fairly recently that studies have been made into how parrots can make the sounds of human speech. Not surprisingly, it was Alex,the famous Grey Parrot, who was the subject of this study. The researchers recorded on video the movements of his beak and, when possible, his tongue; an infra-red camera was used to record his tongue movements and X-ray photography captured the internal movements.

The method of producing words may differ in various species. In the Grey it was found that sound is produced in the

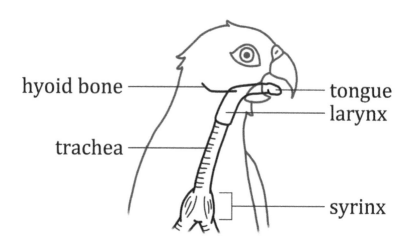

hyoid bone

tongue

larynx

trachea

syrinx

Drawing by Mandy Beekmans

syrinx, the lower part of the larynx, where song is produced. The larynx is the cavity in the throat where, in humans, but not in birds, the vocal chords are situated. A parrot's larynx can only modify sound. A bird's trachea (windpipe) can be stretched or compressed, and these actions are also thought to modify sound. As a parrot's mandibles are hinged there is more movement in the beak than in that of most other birds and this factor is also significant. When Alex made an 'ee' sound his beak was wide open and his tongue was placed at the front of his mouth. When he made an 'ah' sound, his beak was closed and his tongue was at the back of his mouth (Warren *et al,* 1996).

When we speak we do not think about the position of the tongue or the shape of the mouth; we do it instinctively. Obviously our language has derived from the sounds which come naturally to humans. But when one stops to consider parrot mimicry, is it not remarkable that parrots can mimic not only our language, but our method of talking—inasmuch as they can do so given the anatomical differences? The natural language of Greys consists of a series of whistles, plus growls when stressed.

Do parrots understand human speech?

One of the questions asked most frequently about parrots, especially by non-owners, is: 'Does a parrot understand what it is saying?' The simple answer is: 'It can do, if you teach it to understand, just as you would teach a child.' In other words, parrots learn by association. They quickly grasp the fact that 'Hello' is a greeting and that 'Bye-bye' is heard only when someone is leaving or preparing to leave. If you say 'grape' every time you give your parrot a grape, he will soon learn to ask for 'grape' (assuming that he likes them) when you pick up a bunch. Likewise, if you say 'water' every time you refill his water dish, he can ask for water if the dish is empty. But if you teach him to say: 'I can talk, can you fly?' he cannot possibly know what that means. Even a parrot which has been taught the English language over a period of many years—as in the case of Alex—would surely find this impossible. He had been taught to understand nouns and adjectives. In addition, his talking ability clearly indicated he was capable of abstract thought. For example, he could respond to questions regarding the size of objects and state whether they were 'larger' or 'smaller'. But it would be too much to expect that he could have understood an auxiliary verb—one used to form tenses!

It often happens that parrots which are good mimics repeat words which seem an appropriate response to something which someone has just said to them. There are several possible reasons why this happens. One is that the parrot's spoken words were not very clear—and the doting owner has interpreted them in a meaningful way when, in actual fact, they were meaningless. According to some parrot owners, sometimes parrots rearrange words in their vocabulary to produce a meaningful phrase. Such genuine communication might be possible in certain instances relating to very intelligent birds. But birds quite often alter phrases they are taught and, if one happens to make sense, it is more likely to be the result of coincidence.

A story about a cockatoo is told in a very well-known book on parrot care and behaviour. The author was the consultant in a case relating to a cockatoo which was biting unpredictably. The author wrote that she asked the bird what the problem was and he had replied that he was going to get married. The owners said the cockatoo had never been taught the word. The conclusion was that he needed a mate. While Alex has shown that parrots, or some parrots, understand abstract concepts, the reputed response of this cockatoo stretches my imagination far beyond its limits. Even if the cockatoo had heard the word, how could he understand it?

It might just be coincidence when a parrot makes an apparently intelligent response, especially when it can make exclamations which can fit a variety of situations. On the other hand, the person's tone of voice can unwittingly trigger the use of certain expressions, just as a certain tone can trigger laughter, because that tone is heard just before human laughter.

An amusing situation occurred in my own household. When I returned to the UK after living in the Canary Islands, my dog had to go into quarantine for six months. My Amazon had probably enjoyed her absence since she perceived the dog as a rival for my affection and attention. When she came back from the quarantine kennels and I took her into the house for the first time, I picked her up and held her up to my Amazon's cage and told her, 'Look, Lito, Fenny's come back'. My Amazon said: 'Oh, no!' I found this very interesting because she had only ever heard me say 'Oh, no!' when something happened to annoy me. Suppose, for example, I dropped something. My Amazon had lived very closely with me for, at that time, 28 years, and was perfectly capable of understanding when I was mildly annoyed. I cannot recall hearing her use the expression 'Oh, no!' before or since, but, because of the way in which I emphasised Fenny's return, holding the dog up to her eye level, she responded in an equally emphatic way. If I had not drawn to her attention the reappearance of the dog, she would just have eyed her with interest. Instead she apparently responded by showing annoyance in a way that I could understand.

Question
I am very disappointed that my parrot is not learning to talk. Can I use a tape to teach him?

Answer
Playing commercially produced tapes of someone repeating the same phrase over and over will have limited success or, more likely, no success at all. A parrot responds best to the voice of its

owner and to other family members. In addition, listening to these tapes is as boring for a parrot as it is for us. They would surely become no more than a background sound after the first couple of times. They contain nothing to hold the bird's attention. An intelligent parrot like a Grey needs to attach some kind of meaning to a phrase which it hears frequently. This is probably why most parrots so readily learn 'Hello' and 'Bye bye'. They can use them in the correct context—one that they have observed humans using. For a parrot, mimicry is a form of contact with their human 'flock members'. There is nothing to motivate them to respond to a machine.

Taming
To tame a parent-reared or wild-caught parrot usually needs a lot of patience: to force the issue is a mistake. Because most parrots are so possessive of their cages, putting a hand in the cage may provoke an attack. You are an intruder into the parrot's territory. So the parrot must come out of the cage and you must share a small neutral territory. The first step is to persuade the parrot out of the cage. This is easier using a cage with a removable base. Take cage and parrot to a small room, preferably one which is carpeted and has a minimum of furniture. Remove the base of the cage. Either turn the cage on its side so that the parrot can climb out or remove the perch so that it is encouraged to move. When it comes out, remove the cage. Make available a table-top stand with a small ladder or something on which the parrot can perch easily.

Then sit and read or watch television, for at least half an hour, and totally ignore the parrot. The aim is to make him or her realise that you are not a threatening entity. At the end of the allotted time, try to drop the cage over the parrot or to catch it in the least stressful manner possible. Or bring the cage into the room and see if the parrot will climb back into on it its own. This method of taming takes a lot of patience, with apparently little progress at first. But one day there will be a breakthrough.

I well remember using this method, many years ago, to tame a wild-caught Massena's Parrot. One day she flew on to the arm of my chair. I gave no hint that I even knew she was there. After that, taming proceeded rapidly. Once a parrot loses its fear, its curiosity takes over. It will want to nibble at your hand, for example.

Then you are well on the way to establishing a good relationship. But your motto should be to let the parrot come to you. Never pursue it.

Is it possible to tame a screaming, biting parrot which has been locked in a cage for years? An inspiring story in an American magazine for Grey Parrot owners (*The Grey Play Round Table*) shows that in sympathetic hands this can be achieved. He was called Laurie. After his owner died, he had spent six years with her husband—who did not like him. When the old man became too ill to look after him, their son took the Grey to a pet store and tried to give him away. The store owner contacted a parrot hotline who contacted Margaret and asked if she wanted a challenge: a Grey that bit, screamed, swore and plucked. She said yes.

She took him to a vet but he became so frantic with fear that he had to be given oxygen. Margaret took him home and installed him in a spare bedroom. Her teenage daughter would go and sit with him while she did her homework, talking in a quiet voice and avoiding eye contact. Margaret soon began to wonder whether she had bitten off more than she could chew.

> Laurie was a dominant, controlling bird who had been neglected and abused to the point where he was hysterical over the thought of being touched or removed from his cage, had feather plucked himself until he looked like a boiler-fryer with a head, and who was in an extremely poor state of health due to malnutrition, lack of exercise and mental stimulation, and having lived in unsanitary conditions . . . He was also terrified of gloves and was kept in a cage with a padlocked door (Black and Black, 2000).

Laurie had never been taught to obey any commands. After one week training began. His favourite food, peanuts, was removed. He received them only from the hand, in preparation for training. A training stand was made with a perch on the same level as his cage door. If he wanted peanuts he had to come out and take them from the food cup on the stand. After a couple of days, Margaret picked up the stand and whisked him away to the bathroom. She sat down with the stand adjusted so that Laurie's head was slightly above her shoulder.

She could then reach out to him and still have the dominant position of being slightly higher than him—or so she thought. He leapt from the stand. So she sat down on the floor in front of him, talking in a calm, confident voice. She held out her hand. He struck, clawed and screamed—but Margaret would not go away. She gently touched the lower part of his chest while talking to him soothingly. Finally, he threw himself on his back and screamed. It was a heartbreaking response but Margaret knew that she had to continue. If she gave up, he had won the first round and training would be that much more difficult in future. She persevered. He gouged her hands—but within a minute he was sitting on her hand. She talked soothingly while returning him to his cage. He then had a reward of peanuts.

The following day another training session took place. Laurie jumped off his perch every time he was placed back on it. Finally, he stayed there, so he was returned to his cage and given peanuts. On the third and fourth days he learned to 'Step up!' and 'Step off!' and to tolerate having his chest touched. This demonstrates how quickly an intelligent bird can be taught when it realises that it is no longer in charge. It has accepted the trainer as a higher authority. However, he was still rebelling at not being in control. He gave his affection to Margaret's daughter who was not part of the training programme. He would call for her and indicate that he wanted his head scratched through the bars of the cage.

Question
Why does my Cockatiel resist all attempts to tame it? I bought it five months ago in March, as a young bird, but it will not come near me and hisses when I approach. I am very disappointed as my previous Cockatiel was so tame and sweet.

Answer
In the UK Cockatiels start to breed in the spring. As you bought yours in March, it must have been bred in the previous spring or summer, and was too old to tame. It might even have come out of an aviary and is probably very unhappy in a cage. I would suggest you find a new home for it with someone who has an aviary of Cockatiels. You could then buy a young bird from a

breeder. Young ones are available from about May until August, and possibly a little later in the year, depending on the weather. Some breeders sell their young that were not sold at the right age for taming to pet shops where they might spend several months. Most pet stores will sell these Cockatiels and tell the purchaser that "it will soon become tame". You should only buy a Cockatiel in nest feather and one that appears unafraid if you want a good companion. The best time to purchase one is within one month of it leaving the nest. This advice applies to all parrots but if they are seasonal breeders, such as Cockatiels and Amazons parrots, do not buy one between November and about May because it has almost certainly passed the stage when it will be easy to tame.

TCM—Traditional Chinese Medicine

Question
Why does my parrot like to have her feet rubbed.

Answer
A German veterinarian, Dr Rosina Sonnenschmidt, developed a technique which she called acupressure tickling. Knowledge of the avian energy points should be an essential part of bird keeping, she wrote. For example, massaging the underside of the ball of the foot has the effect of toning and calming the metabolism. This is important because many diseases start with a blockage of energy in the kidneys. She wrote that almost all kinds of feather plucking, infertility, anxiety and liver problems stem from low kidney energy. My Amazon, age unknown, is not in perfect health. Her last vet check included an X-ray which revealed slightly enlarged kidneys and a blood test which indicated slightly elevated liver enzyme levels. I was now beginning to understand why she enjoyed the sensation of having her feet massaged.

Dr Sonnenschmidt has saved birds from death simply by acupressing certain revival points, and had instructed others how to do the same. On one occasion she received an urgent telephone call from a lady whose Umbrella Cockatoo had suddenly collapsed. It was a very aggressive bird and not tame. Normally it attacked anyone who entered the aviary. Dr Sonnenschmidt faxed its owner a diagram showing the survival points. This is what happened:

> The moment she touched the kidney point under the foot, the Cockatoo relaxed and thoroughly enjoyed the acupressing of all points. When the (trembling) lady stopped the treatment, the bird clearly signalled it wanted the massage to continue. This wild and aggressive bird understood that the fingers of its owner were not enemies at that moment but a source of energy! (Sonnenschmidt, 1996b).

Television

Television can be both harmful and beneficial to a companion parrot. It would be harmful if the bird were kept near a TV set which was switched on for hours on end, with its flashing lights and usually loud sounds. For short periods exposure to television can be beneficial.

There is no doubt that some parrots greatly enjoy watching television, whereas others ignore it. Many are interested in birds and animals on the screen, especially parrots. To amuse a wild-caught Blue-fronted Amazon who was in my care for a few weeks, I would show him DVDs of parrots in the wild. He watched these avidly and never tired of them. The first time he saw a sequence of Amazon chicks in a wild nest, he became so excited that I was convinced he must have reared chicks before he was trapped. (It made me sad that for the 20 years or more since then he had lived a solitary life.) Many Amazons also enjoy pop music programmes because they find loud music stimulating.

If your parrot is a good mimic, it is not a good idea to let him watch programmes which consistently feature bad language. Many parrot owners claim that their birds have picked up new words from the television. Other parrots learn to whistle popular signature tunes.

Many parrot breeders use observation cameras and monitors to watch what is going on in their aviaries, also for security purposes. Small cameras within nest-boxes are now widely used. They provide a fascinating insight into a world which was a closed book only a few years ago. I have an observation monitor which shows

Well, my parrot is always on television!

me what is going on in a couple of nest-boxes of small lorikeets. Because I prefer to have a close-up view of the part of the box where eggs are laid, rather than a wide-angle view of the nest-box, only a part of a lorikeet's body is in view at any time.

The monitor was quite close to my Amazon's cage. When it was first installed one of the lorikeet pairs was preparing the nest—chewing up the wood shavings and scratching about in them. My Amazon appeared to watch this with interest, but as the bird's head was not usually in view—mainly the body—at first I was not sure whether she could interpret what she was seeing, especially as the picture is in black and white. Three or four days later I was left in no doubt that she did understand what she was watching. She went down on the cage floor and started to chew up newspaper. Watching the pair of lorikeets had apparently stimulated her. Thenceforth I turned the monitor so that it faced away from her cage—and the paper-shredding stopped.

Question
My Blue-fronted Amazon seems to enjoy watching television. Certain things seem to catch his attention. I am curious to know what he sees. Does he see the picture exactly as we do? Sometimes he seems to be reacting to something on the screen, yet only one eye is facing the TV.

Answer
This is a very interesting question No: he does not see the picture exactly as we do. Only some birds of prey and owls see an object with both eyes as we do. The eyes of other birds are placed so far back on the side of their head that the two fields of vision overlap only slightly. If a parrot wishes to see directly in front, it must turn its head slightly. A bird's eye is much larger in proportion to the size of its head than that of a human. In fact, the weight of birds' eyes is about 10–15 per cent of that of the head; in humans it is only 1 per cent. The eye structure gives a parrot a wider field of vision, and the extent of the back of the eye gives a large picture. A bird's eye excels over any other vertebrate's. (But note that parrots of red-eyed mutations do not have such good sight as other parrots.) Birds can distinguish smaller objects at a greater distance much more clearly than we can. I sometimes observe my aviary birds looking high into the sky with alarm. I am not able to see what it is that has frightened them.

Territoriality
Mature parrots, especially males, can become extremely aggressive protecting the area which they consider to be their exclusive territory. This is usually the area around their cage but, when they are in breeding condition, they might defend a location which they perceive as a nesting site. This is usually a secluded dimly lit place. It could be an open drawer or cupboard or the folds of floor-length curtains where they reach the floor. Wherever it is, it can become a danger area for humans living in the house. An unprovoked attack may occur on the unsuspecting person who enters this territory. Or attacks might

be commonplace in the area around the parrot's cage. This will certainly remove the enjoyment from letting the bird out and might even lead to it being found a new home.

But there is an answer—re-training—or training, as the case may be. Discipline is needed to bring him under control again. It may be a temporary problem due to hormonal influences or it may be a permanent one due to the fact that the parrot sees itself as the dominant member of the household. In either case it needs to be taught or re-trained the 'Step up!' and 'Go down!' commands. This training must initially take place in a room away from the cage, using a play stand or portable perch. Gradually, when the parrot is responding well to the commands, the perch should be moved nearer and nearer the room containing the cage, then inside the room, with the training continuing. Thenceforth the parrot should always be returned to the cage with the 'Go down!' command and removed from it with the 'Step up!' command. The cage door should never be left open for him to emerge when he wishes, as this just reinforces the idea that he is in control. Aggression almost always subsides when the parrot realises that this is no longer the case.

Even the tamest and sweetest-natured parrot can be very territorial regarding the interior of its cage. This is one reason why it is so important to teach a parrot to step on to the hand. Obeying this instruction most parrots will step on to a hand placed inside the cage. If they are not trained to do so they are likely to attack the hand as an intruding object. Owners of more than one pet bird should never allow one to land on the top of another's cage. Even if the two birds agree in neutral territory, the top of a cage with the owner inside it is a danger zone. One unfortunate little parrotlet lost its left leg when it landed on the top of an Amazon's cage. The parrotlet had been left unsupervised. This is the kind of accident which could be predicted almost with certainty by someone who has had experience of Amazon Parrots.

Question
Three weeks ago I bought an Umbrella Cockatoo from a pet shop. He had had at least one previous home. He seemed very well behaved and not too noisy. However, during the past week he has become extremely loud, yelling and screaming for long periods. Why has his behaviour suddenly changed?

Answer

When a parrot is removed from its usual environment, it is normal for it to be quite subdued for a period of about two weeks. It has been displaced from its established territory; everything around it, including the people, are unfamiliar. As it settles in, it begins to acquire territorial instincts regarding its immediate environment. Parrots advertise the fact that a given location is their territory by making loud vocalisations and, if necessary, using aggressive behaviour towards those who enter the territory. In other words, your cockatoo's behaviour has now reverted to normal. Cockatoos are naturally very noisy birds— a fact which should be understood by every prospective purchaser. If anyone is told of a cockatoo: 'It's a very quiet bird,' they should suspect that something is wrong with it.

Territoriality in the aviary

It is extremely important that keepers of aviary birds understand the territorial behaviour of parrots. Countless birds have died because they didn't. If this paragraph saves even one life, this book will have been worth while. A parrot defends its territory against all comers. If one member of a pair dies and the owner obtains a replacement, he or she should never place it in the aviary inhabited by the survivor. It is likely to be attacked. The best procedure is to place the new bird in an aviary on its own for a couple of days, so that it can settle down unmolested, then place the potential mate in the aviary. If a spare aviary is not available, the original inhabitant should be put in a cage within the aviary or shelter. When the new bird seems confident, the other one can be let out. This procedure should also be adopted when the member of a pair which is not the dominant bird has been sick and removed from the aviary for a period. Even a member of a long-established pair might be attacked when it is returned to the aviary, if this precaution is not taken. Some species are much more territorial than others.

Toenails

Toenails receive mention because they could affect behaviour, including that of the person handling the bird. If a parrot's nails are very sharp the back of the handler's hands will become covered in scratches, making handling an unpleasant experience. Parrot's nails are like our own; they are made of keratin and grow continuously. Normal usage wears them down. However, this will not occur when a parrot is kept on the wrong surface, such as plastic or other smooth perches, or if the parrot sits for hours on the bars on top of his cage.

The best surfaces are:

1. Freshly cut bark-covered branches of medium-hard wood such as apple and other fruit trees. The bark will soon be removed. The perch must be replaced before it becomes slippery and shiny with wear.
2. Orthopaedic perches (and stands with these perches) made for parrots, obtainable from some pet shops. Care should be taken with concrete perches which can be purchased in a range of colours. Some of these are so rough that they could damage the tender skin on the underside of the feet.

Parrots should have access to perches of a variety of different thicknesses, to give foot exercise. They will often chose thinner perches than one would expect. Note that vertical perches are enjoyed by some birds and will help to strengthen their leg muscles.

The need for regular nail cutting (except in old birds) is an indication that the parrot is not being kept on the correct surface. If nail cutting is necessary, the bird should be held firmly in a towel; a second person should attend to the nails. Some young parrots have very sharp nails which should be filed—nail cutting, in this instance, is not recommended as it may promote nail growth, resulting in regular cutting—a stressful experience for most parrots. If the nails become overgrown to the degree that normal wear will not rectify the problem, then nail cutting is necessary. A vein carrying blood extends nearly to the tip of the nail. If the vein is cut, profuse bleeding will occur and the sore nail tip will cause immense pain and discomfort. Excessive bleeding can endanger life and must be stopped. (In an emergency hold the parrot's toe in flour.)

If a parrot's nails are too long they will catch in the wire of a welded mesh cage, making climbing awkward and stressful. Long nails also make it difficult for a parrot to grip the perch efficiently. This could cause reluctance to move around and explore. In some old parrots the nails grow in an irregular shape, even in a corkscrew. They must be attended to by an avian vet if necessary.

Question
Why does my parrot fall off his perch after his nails have been cut?

Answer
Nail length is one of the factors which affect the way in which a parrot grips the perch. It will need to readjust that grip when suddenly all its nails have been shortened. The adjustment will be made quite quickly, provided that the nails have not been cut too short.

Question
Why does my parrot bite her nails?

Answer
Parrots occasionally nibble at the nail tip to keep it in good condition. However, if this action is repeated often, like a bad habit, it could indicate that the parrot is anxious or stressed.

Toes, mutilation of

If a parrot starts to bite at its toes—not merely just removing dead skin—immediate veterinary advice should be sought because it can lead to self-amputation. It is a very distressing and puzzling condition, in which parrots have been known to bite off their toes, presumably due to severe pain. When I tried to advise a friend whose Meyer's Parrot was biting off its toes I sought the advice from an avian vet in Australia, Stacey Gelis. He told me that this has occurred in a broad range of parrots. In his practice underlying disease was identified or suspected through further work such as blood tests and radiographs.

His suspicion is that the birds suffer from a neuropathy (nerve irritation) which causes a "pins and needles" sensation in the offending limb and results in self mutilation. Some of the more frequently diagnosed causes include heavy metal toxicity (especially lead), kidney disease leading to sciatic nerve irritation and sometimes underlying bacterial/viral diseases including chlamydiosis and PDD (psittacine dilatation disease). Ergot (fungal) poisoning can also lead to necrosis of the ends of the toes which may result in self mutilation.

There are also cases which seem to have no identifiable underlying cause which are treated with drugs such as gabapentin which help in cases of nerve-related pain. Even in these cases an underlying disease process is suspected, though not diagnosed, rather than a purely behavioural cause.

Tool-use

Tool-using has been recorded in only a small number of birds and mammals, notably chimpanzees, parrots and members of the crow family. It is generally considered to be a sign of intelligence. However, tool-using also occurs in, for example, the Woodpecker Finch of the Galapagos which feeds on the grubs of wood-boring beetles which it reaches by digging beneath the bark with its sharp beak. Some of the grubs are out of reach so it breaks off a cactus spine, probes into the hole, spears the grub and pulls it out. Although some parrots in the wild feed on grubs or other insects, there is no record of them behaving in this way or of using tools for a useful purpose. But tool-using is included here because it has been recorded by countless parrot owners. The dexterity of beak and foot in parrots, combined with a high degree of intelligence and playfulness, results in parrots using objects for various purposes—but usually for play.

The behaviour of a group of Goffin's Cockatoos monitored by researchers in Vienna has already been mentioned (see Intelligence). In 2012 one of these birds featured in all kinds of media news stories, including clips on the internet, to demonstrate how clever a bird can be! Named Figaro, he learned how to retrieve items dropped through the welded mesh of his aviary to the outside. He would bite off a splinter of the wood that fronted the aviary, push it through the mesh and rake in the food item.

Two years later the researchers experimented with six Goffin's Cockatoos, three males and three females, who had watched him perform. Interestingly, the males also learned to do this although their technique was not necessarily the same. Sometimes a cockatoo would lose the wooden tool he had been given, leaving it out of reach. In that case, he would reach for another tool but would not use it to retrieve the food—instead, he would retrieve the first tool, then use that to get the food. The females did not learn tool use.

The most simple form of tool created by a parrot is when it sharpens a small twig to a satisfying point, then uses it to scratch its head feathers. Birds kept alone have no mate to preen their feathers—so they use their ingenuity. It seems unlikely that a bird with a partner would do this.

Try giving your parrot a hard plastic bottle top. It will probably try to destroy it but it might put it to another use. Many parrots use such items as scoops and amuse themselves for minutes at a time scooping up seed or water. Why do they do this? The simple answer is that they are playing—just amusing themselves, as a child might. I have seen parrots use a half walnut shell in exactly the same way.

Touching

Many parrots make very affectionate pets who love to have their heads scratched and who return affection, by preening the hair or around the eyes of their human friend. Most parrots object to being touched in areas other than the head, especially to being touched on the tail. Cockatoos are an exception. They like to be caressed all over. However, it must be realised that other parrots do not normally tolerate such familiarity. Furthermore, some parrots do not want to be touched at all. No matter how tame they might be in other respects, they do not welcome the human touch.

In some cases there is no explanation for this other than the personality of the individual bird. On the other hand, the reason might be because in the past someone has moved a hand too close and too quickly on repeated occasions, thus frightening the parrot. If it knows the intention, such approaches might be more acceptable. Prefacing an attempt with the word 'Touch', or another chosen word, consistently used, might lead to success, especially in birds whose aversion to the human touch stems from the fact that they were not touched when young. However, if in the long term nothing will change the bird's aversion to being touched, the parrot can still be an adored companion. The humans around it must accept that touching is not permitted—and make this very clear to visitors.

The initial attempts to touch a parrot should consist of scratching the nape feathers, as this is the area most universally accepted. Actual stroking should never be carried out. There are two reasons for this. The often clammy human hand can disturb the feathers and, if not clean, could even transmit disease. If a parrot is consistently stroked in the same place by someone with nicotine on their hands, or by someone who perhaps sits eating crisps (potato chips) while watching TV and stroking the parrot, the build-up of oil or nicotine could even cause the parrot to pluck feathers in that area.

The second reason is that stroking a female parrot on the back resembles the action of the male mounting the female for copulation. In some females stroking the back will initiate the crouching stance with which she invites a male to copulate.

Toys

Since the 1990s the change in attitude towards giving toys to parrots has been remarkable. It is part of a new understanding of the needs of companion birds.

They *must* have items with which to keep their beaks and minds occupied. Formerly, if a parrot received a wooden cotton-reel to gnaw on, it was a lucky bird. These days some parrot owners spend a lot on expensive and intricate toys. This is not necessary (but it does show that many owners realise how important it is to keep their parrots amused). The success of a toy can be judged not on how good it looks but how long it will keep a parrot busy. The other aspect which must be considered is safety. A toy with open links, destructible plastic or made with some forms of galvanised wire is potentially dangerous or even lethal.

Note that galvanised metal which has been electroplated is safe but that zinc might be ingested from hot-dipped galvanised wire. Now that veterinarians know how to detect heavy-metal poisoning, greatly increased incidences are being reported. If your parrot is lethargic or sick and the usual tests are inconclusive, ask your vet to test for metal poisoning. Don't let your parrot play with keys or other potentially poisonous metal items. One cockatoo was diagnosed with zinc poisoning and the source was a mystery. Finally, the owner realised that it came from a silver neck chain worn by her husband. The cockatoo

Great! a new box

loved to play with it. The chain was made from silver plate over zinc.

For full-winged parrots, toys which permit and encourage vigorous exercise are worth fixing up. For example, a swing can be hung from the ceiling using cotton rope. Such items will improve a parrot's landing skills and aerial manoeuvrability, at the same time giving great enjoyment to parrot and observers. A rope hanging vertically from the ceiling is equally effective. Toys can be made using offcuts of hard wood threaded on to leather. These are totally safe and fulfil their urge and need to gnaw (*see* Destructiveness.) Small toys of this nature should be offered initially. If they are too large they might be viewed with alarm.

Training

There are two important facts to bear in mind about training a parrot. One is that it takes time, kindness and patience. The action or behaviour being taught may need to be reinforced by repetition occasionally. The second point is that many people are unconsciously negative—whether they are training a dog, a child or a parrot. Yelling 'Don't do that!' might have some effect on a dog or a child, who will at least understand your intention from your tone of voice—as indeed will most parrots. But raising your voice to a parrot will have one of two detrimental results:

- The immediate and excited response it receives, perhaps, as a result of yelling or biting, is perceived as a reward. It has grabbed your attention. This is perhaps all that a tame bird wants.
- On the other hand, a nervous parrot will respond very badly to glares and a loud voice. It will do nothing to help you build a good relationship with it.

Training is important for two reasons. It enables you to control your parrot's movements when he is out of the cage, with 'Step up!', 'Go down!' or other instructions. There is nothing worse than having to chase a bird (stressful to both parties) around a room in order to return it to its cage. Secondly, training imposes discipline. Without discipline, there can be little or no respect for you from your parrot. Training will win that respect—not by domination but with the knowledge that there are rules which must be obeyed. It is no different to the situation which exists between parent and child but, in this case, it is the flock leader (you) and your parrot.

Training should be carried out in a quiet place where there are no distractions. The room should be as small as possible with little furniture so that your parrot is easily retrieved. If he is wing-clipped, the area should be carpeted so that he cannot hurt himself if he falls. No other people should enter during the training period. During training sessions you must be calm and gentle. If you are not in a sympathetic state of mind (perhaps after a bad day at work), postpone the session until you are more composed. Your agitation would be 'read' by your parrot. Whatever you are teaching him, make a mental picture of this, with him reacting as you hope, and in a confident manner. Conjure up this picture every time. Not only does it help you to be more positive, but the possibility exists that your parrot can tune in to this picture. Training sessions should last between about five and 15 minutes—or they should cease as soon as the bird loses interest.

The 'step up' command

The most important command that enables a person to handle and move a parrot is usually known as the 'Step up!' command, i.e., it teaches a parrot to step on to the offered hand. In a small parrot, such as a Budgerigar, it is known as finger-taming.

The person training must, of course, be comfortable handling the bird. This will not be the case if its nails are very sharp, as is the case in a young Grey, for example. It is then best to file the nail tips using a nail-file or emery board. They need to be only slightly blunted. (Cutting may encourage nail growth, which should be avoided—(*see* Toenails.)

Training a young hand-reared parrot to step up is easy, as he has been used to being handled. However, as a parrot usually likes to be at the highest point, his natural reaction after stepping on to the hand is to climb up the arm and sit on the shoulder. This must not be permitted. Keep the wrist elevated and held slightly away from the body to curb the inclination. When you initially offer your hand say 'Up!' When he stays on your hand keep his attention by praising him with 'Good boy!' If necessary, as soon as he steps up, use the free hand to make him step up again, then put the parrot down on another surface. Even when this command has been learned, always reinforce it by using it several times a day. If possible, end every session with a small success or on a positive note.

Teaching a parrot which has not been hand-reared or one which is no longer easily handled is much more difficult. First, one has to win the bird's confidence. To start the 'Step up!' training before this occurs will do more harm than good. It will frighten him. If a parrot is not tame enough to let out of the cage, much time must be spent sitting talking to him, and offering a favourite tit-bit, through the cage bars. When this is regularly accepted, the next move is to open the cage door and offer it. Keep this up for some days then, instead of offering a tit-bit, move you hand slowly towards his breast and push gently, just above the legs, talking softly all the time. If he tolerates this, offer the tit-bit afterwards. If he does not, try again, a few hours later. It could be days or even weeks before you succeed—but persevere.

When you do succeed, do not try to take him out at first. Just praise him for stepping on your hand. When you feel he is confident about this and your command 'Step up!' is succeeding on most occasions, you can take him out of the cage. Be sure always to use the same command words and to praise good behaviour in a positive and happy tone of voice. You might even want to add a little whistle as many parrots love the sound of whistling.

What do you do if the attempted training is a total failure because your parrot reaches down and tries to bite you every time you put your hand in the cage? There are at least two possible reasons for this. One is that your movements are so hesitant that your parrot is unclear about your intentions. You must move your hand steadily but not too slowly towards his lower breast. If you are really nervous about being bitten, you could try covering your hand with a piece of towel as near to flesh colour as possible, to give you more confidence. Leave it near the cage before you use it, or this strange material may frighten some birds. However, training is less likely to succeed by someone who is frightened of being bitten. The bird can sense that and knows that the person can easily be intimidated. It will therefore not respond to attempts to train it.

The second reason why your parrot attempts to bite is because he is afraid and not yet ready for training. You must do more work to win his trust. You should always talk quietly when approaching him. Silence can make a parrot suspicious and fearful. To a parrot, vocal communication is so important.

A further reason why a parrot, especially a young bird, tries to bite during 'step up' training is that he is testing you—trying to discover whether he can dominate. The best reaction is to ignore the bite totally and try to stop yourself from crying out. If a bite brings no reaction, the parrot should normally soon cease from attempting this. If it brings reaction in the form of loud cries, he may think he is initiating an exciting game that is sure to bring a response from you. If the handler receives a painful bite, it is difficult to carry on as though nothing has happened. In these circumstances it is probably wisest to cease the training session—but without a word or a look. Just stop if the bird is inside the cage or put him away if he is outside. It is no good carrying on when your frame of mind is no longer sympathetic.

At this point it is important not to chase him around. If he does not respond to your 'Step up!' command, how do you get him back into the cage? The first time you let him out should be in the evening. You can thus note his location, quickly turn out the light, cover him in a small towel and quickly transfer him to his cage. An alternative idea with a small parrot species or Budgerigar, is to use a cage with a detachable plastic base. Before you turn out the light, detach the base from the top part of the cage. In the darkness you can drop the top part of the cage over him.

The basic 'Step up!' command should ultimately be useful for retrieving your parrot and replacing him in his cage. However, if he is full-winged he may have other ideas. Just as a child does not want to be called in from play, many normally obedient parrots avoid being returned to the cage by flying off. Chasing a parrot around a room is a frustrating and stressful experience which should be avoided. Also it undermines the control which you otherwise exert and allows the parrot to believe that he can outwit you. As he is the possessor of wings one has to admit that this is true. Nevertheless, in my opinion, this is not reason enough to clip a parrot's flight feathers. Another way must be found.

A friend whose Red-bellied Parrot objects to being returned to the cage has found a stress-free way of doing this. When she is perched on his hand, he scratches her head while moving towards the cage. His hand is cupped over her small body, without actually touching it, as he scratches her head. By the time he reaches the cage, it is an easy matter to keep scratching her head as he places her inside. Note that if an attempt is made to handle a parrot bodily at this time, it would almost certainly attempt to bite. Other ways can be found to entice a parrot back to his cage, such as only placing the most favoured food item in the food cup when it is time for him to go back. In all cases, gentle persuasion must be used. If the problem is a permanent and tedious one, more training is needed.

Trying to teach some parrots to step on to the hand is difficult or impossible because they are afraid of hands. In this case, teach them to step on to a short perch, such as a piece of dowel. (*See also* Hands.) This is a good method with adult cockatoos and macaws, many of whom will initially step on to a perch much more readily than on to a hand.

While the 'Step up!' instruction is being mastered by the parrot, training can start on 'Go down'. This is easier as it only involves using these words as you put him back in his cage, or place him on another surface when he is out of the cage. It will not be long before the use of these words has the desired action.

So what else can be taught? Right from the start it should be made clear to your parrot that you do not want him to land on your head? Why not? He is (or will be, when he is older) difficult to retrieve there and may bite at the approaching hand. Also, you cannot see what he is doing or watch his eyes, which indicate what sort of mood he is in. Again, this increases the risk of being bitten. If your parrot is full-winged, you just duck when he comes in to land on your head. He will soon get the message. If he is clipped, never permit him to climb on to your head from furniture or shoulder.

Some behaviourists advocate using the 'OK!' message. This word is used, they suggest, when your parrot has obviously decided he will do something and you know you cannot stop him. By saying 'OK!' you are giving him permission, inferring that you took the decision away from him. I see this as a double-edged sword. Are you going to say 'OK!' as his beak connects with your expensive new sofa? In any case, by the time you have interpreted his impending action, he has already decided to do it. Some birds, especially Greys, will look at you slyly just before they are about to do something which they know is forbidden. Surely, this is the time to say 'No!' and point your finger. If you say 'OK!', how can you teach your parrot not to do it again? A young bird might look to you for guidance when about to do something about which he is uncertain, but an adult will only look at you to see if you are watching him. Can he get away with it?

When a parrot responds to one or more commands (especially 'Step up!') he will realise that you are a higher ranking flock member than he is. Life will then be much easier.

A different kind of training is, in fact, teaching a parrot about the objects in its environment. The most important of these is window glass. Once your parrot steps up on command, you should take him over to a window and gently move your hand forward until his beak is touching the glass. Do this several times and at different windows. This lesson is quickly learned.

Sometimes we forget that parrots evolved in an environment which is in total contrast to the artificial confines of our living rooms. They need help to survive in our houses, whether or not they are full-winged. One way in which we can help them is to move around the house, parrot on hand, showing them things. We should never allow them access to potentially dangerous objects. Parrots should never be allowed to land on gas or electric cookers. Obviously, if they landed there when the cooking rings were hot they could be terribly injured. This is where the 'No!' command is useful. If a parrot lands on such an appliance, it should be removed instantly with a stern 'No!' But it is not the parrot's fault if a person is foolish enough to let it out during (or soon after) cooking is in progress.

Question
I have been given a Grey Parrot which has not been handled for more than ten years, since its original owner died. Apparently it was once very tame but now seems rather suspicious. Will it be possible to train it to come on my hand?

Answer
Yes, if you have the confidence, patience and the ability to win the trust of your Grey. Unlike some parrots, Greys can be retrained in a surprisingly short time—in sympathetic hands. So much depends on the attitude of the person undertaking the training. If he or she fears being bitten, the parrot will sense this, making training so much more difficult. It often happens with parrots which have not been handled for years that they warded off human attention by biting. Thus they perceive themselves as being superior to people because they can get rid of them so easily. You must therefore stand your ground, even if you receive a couple of bites initially.

Tricks
Parrots can be taught to do a variety of tricks. While some people might think this is a way to strip a bird of its dignity, in fact many parrots love to perform. They would not do so otherwise. Teaching tricks is actually highly beneficial for the bird because (a) any action taught and performed on command is a kind of discipline, and (b) the attention necessary to teach any trick will be enjoyed by the parrot and helps to break the monotony of the day. Your parrot will also relish being the centre of attention when you show off his accomplishments to friends.

The parrot's ability to learn tricks was one of the factors which endeared it to early pet keepers. The accomplishments of a Senegal Parrot called Joey were described in a book published in 1928. Remember that there were no captive-bred Senegals available then. The bird would have been wild-caught.

> . . . the keeper says 'one' and the bird stands to attention; 'two' and Joey stiffens; 'three—now die' and slowly the parrot sinks on to his back, clenches his feet and lets his head sag backwards over the edge of the man's hand. He looks exactly like a dead bird, and he remains 'dead' until told to come to life again. His movements are always leisurely whether dying or reviving, and he will repeat the performance again and again . . . (Sidebotham, 1928).

One of the most remarkable performing birds on record was an Indian Ringneck Parrakeet. It was bought in Calcutta by Alfred Ezra in 1910. An Indian had taught him his tricks which he would perform in the presence of huge crowds for money. His accomplishments included twirling a lighted torch—a long stick, twice as long as the Ringneck, which was alight at both ends; drawing water in a small wooden bucket; retrieving anything thrown to the other end of the room; loading and firing a little cannon, which went off with a loud noise. His most difficult trick was to thread beads. A few glass beads were placed on a table with a piece of thread with a knot at one end and a short blunt needle at the other end. He thread the beads using his tongue and beak. Alfred Ezra recorded: '. . . only kindness and coaxing with a grape or a nut used to make him do his tricks quickly. Once when in a hurry, I was rather impatient with him, with the result that he refused to do anything, and flew away each time I brought him to the table' (Ezra, 1929).

Some bird parks and zoos still have shows in which parrots do such tricks as riding a bike, pulling a cart and fitting shapes into slots of the right shape on a board. In fact, parrots are capable of learning far more complicated actions. The teaching method is usually food deprivation and sunflower seed rewards. I think food deprivation, if it results in periods of hunger, is wrong and that many parrots could be taught using head scratching and caresses in place of nuts or sunflower seeds.

Question
Can you explain how I can teach my Blue and Yellow Macaw to shake hands?

Answer
When your macaw is perched on your hand, hold the first two fingers of your other hand just above one foot. (Note whether your bird is right-footed or left-footed and hold your fingers above the preferred foot.) To prevent him from stepping up, do not hold your hand in the way you do when you want him to step on—present your fingers and tuck the other fingers and thumb into the palm. Give the instruction: 'Shake hands!' At first he will not, of course, know what this means. He might try to step up. Try to encourage him to hold the end of your finger with his foot. If he does so, gently and slowly move your finger up and down, repeating 'Shake hands!' Praise him all the time he manages to do this. At first he will be very uncertain of what is expected of him, so keep praising him. When he has mastered this, if he is adult, don't expect him to do this for other people, because if they do not approach him in a positive manner, they may get bitten.

V

Veterinary surgeons

Question
Why does my vet know nothing about parrots?

Answer
In the five-year training period of a veterinarian, only a couple of hours are devoted to birds other than poultry. The few avian vets in the UK have gained their knowledge on the treatment of parrots and other exotic birds by encouraging clients who keep them, by attending conventions worldwide for avian veterinarians and by contacting other avian vets through the Internet. In other words, they have worked hard to gain specialist knowledge. Even though it will usually mean travelling a considerable distance to seek out an avian vet, this will usually be more satisfactory than consulting a nonspecialist. Many vets who are not used to treating parrots do not know how to handle them—and some have no wish to learn. Worldwide, some specialist magazines publish lists of avian vets.

Vision, colour

Birds see colours differently. They can see more clearly at a distance because their eyes are not very sensitive to the blue light rays which make distant objects look hazy to us. They are more sensitive to the red rays than we are. Red gives them the most illumination (Hess, 1951). This is probably why so many berries are red when ripe and why so many flowers in the tropics have red blooms where there are nectar-feeding species. Birds are attracted by this colour.

Birds do not see colours in the same way as do humans. When birds are exposed to ultra-violet light—either through sunlight or through a UV lamp made for captive birds—they are able to see a wider range of colours. The avian eye possesses four cones—one more than the human eye and it is the fourth cone that gives this advantage. Wild parrots are thus able to detect the degree of ripeness of fruits. However, many fruits are eaten before they are ripe. Fruits that reflect UV light are more conspicuous to birds against green leaves, whether in sun or in shade. On Barro Colorado Island in Panama scientists found that birds removed nearly all the fruits from plants in UV light but removed fewer than two-fifths from fruits where this light failed to penetrate.

Question
Is it true that the plumage of parrots looks different to parrots than to humans?

Answer
Yes, this is true for some species. Areas of plumage of some parrots, especially yellow, fluoresce under UV light. For example, this applies to the Golden or Queen of Bavaria's Conure (*Guaruba guarouba*) which is nearly all bright yellow with some green wing feathers. Under UV light a square patch on the nape fluoresces yet to the human eye it is uniformly yellow.

W

Wheezing

Question
My pair of African Grey Parrots have lived in an outdoor aviary during the 18 months I have had them. They are very easily frightened and I assume that they were wild-caught birds. When I go near their aviary one of them starts breathing rapidly and wheezing. I can see its throat moving in and out. As soon as I leave the aviary, after feeding them for example, it stops wheezing. Is this a behaviour problem or is it caused by disease?

Answer
Wheezing is a sign of extreme stress which is most likely to be seen in species of a nervous temperament, such as Greys and *Pionus* Parrots, especially in wild-caught birds. In *Pionus*, the affected bird is obviously stressed, and its whole body is heaving with fear. As your Grey's wheezing ceases so soon after you leave the aviary, it is unlikely to be caused by disease. I would suggest that you use swing feeders or have a feeding hatch, so that entering the aviary is kept to a minimum and is not a daily occurrence. Wheezing in Grey Parrots might also be due to respiratory problems, such as sinusitis, caused by a Vitamin A deficiency. But in the case of your Grey, it is almost certainly due to stress. In the case of a respiratory condition, breathing is usually laboured after flight.

Whining

In cockatoos whining and swaying is a sign of forced-weaning (weaning too early) or weaning without emotional support.

Weaning relates not only to food independence but to the bird being ready to break free of the bond which has tied it to the hand-feeder. In the wild the larger cockatoos (not Galahs or Bare-eyed Cockatoos), but certainly the black cockatoos and, most probably, Moluccan and Umbrella Cockatoos, stay with their parents for many months, up to nearly one year. No research has been carried out on Moluccan and Umbrella Cockatoos in the wild but these species, along with the Black Cockatoos, require the longest weaning periods. In a captive situation, parrots hand-reared with siblings, or with the same or related species, generally wean more quickly than a parrot which has been hand-reared on its own. The latter can take months longer than normal to become independent. Moluccan and Umbrella Cockatoos hand-reared with others of their own kind can, in my experience, normally be considered weaned between the ages of five and six months. Attempts to wean them at 14 weeks are cruel—and will almost certainly result in prolonged or life-long psychological problems of which whining is just one symptom.

Wing-clipping

Wing-clipping is one of the most emotive and controversial subjects in the world of companion parrots. I am adamant in my dislike of this practice. In my view wing-clipping is not only the equivalent of mutilation and a form of psychological abuse, but it is often carried out in order that a human can dominate a parrot. It also results in serious injuries.

Many owners of clipped parrots will take exception to my views. The usual argument is that a parrot's wings must be clipped for its own safety. The way I see it is that a wing-clipped parrot can be kept in a household where this would otherwise be impossible—either because of the danger of escape or because most people do not know how to control or discipline a parrot which is not wing-clipped. My answer to this in both instances is that a parrot is not a suitable pet for that particular household. This point can be illustrated by a sad story which appeared in a bird magazine. A pet cockatoo escaped. It was several days before the bird was located. The owner's vet advised her to clip the cockatoo's wings. She did this with great reluctance. A few months later the cockatoo

escaped again. It went out into the road where it was immediately killed by a car. What was needed, as is usually the case,
was not wing-clipping but better vigilance on the part of the
owner.

Wing-clipping is a comparatively recent innovation in parrot
keeping. The blame must be laid squarely at the door of some
breeders and pet shops, who do not even give purchasers the
option of a full-winged bird. I know of people in the UK who
have ordered young parrots, waited weeks for them to be ready,
to find to their dismay that the wings had been clipped.
Prospective purchasers should clarify this important point in
advance. They should refuse to accept a wing-clipped bird
unless this was requested.

In the USA wing-clipping is standard practice to the degree
that it is assumed that a newly purchased, recently weaned,
hand-reared parrot will have had its wings clipped. When I
started keeping parrots it was almost unknown to wing-clip
those kept as pets, with the exception of the large macaws.
Many of my friends had full-winged parrots whizzing around
their houses (as, indeed, I did) and even their gardens. There
is nothing to compare with the joy of watching a parrot fly,
even within the house. To my mind, to deprive a bird permanently of its power of flight is an unacceptable form of
cruelty.

A parrot's body and plumage is built for flight. Flight is in the
very psyche of the bird. To deprive it of this fundamental act
must have consequences which it is difficult for a mere human
to perceive.

Let us examine the arguments in favour of wing-clipping.

- *Clipping the wings prevents a parrot from escaping.* In fact, as
  many or more of the countless parrots which escape have
  had their wings clipped as those which are full-winged.
  Owners of full-winged parrots are ever mindful of the haz-
  ards presented by an open window or door. Many owners of
  clipped parrots are careless and do not notice when clipped
  flight feathers have been replaced by new ones. I have even
  heard of owners of clipped parrots who were under the
  impression that once the flight feathers had been clipped the
  parrot would never fly again. Many parrots which have been

Don't let them see you swimming, they'll clip your fins

clipped for years do not attempt to fly. However, if some of their flight feathers had grown back they would instinctively take off if suddenly alarmed.

- *Clipping keeps a parrot safe from mischief it might get into if it was full-winged.* But it also makes it vulnerable to attack from cats, dogs and other animals. If it tries to escape, it would be unable to fly.

- *Wing-clipped parrots can be allowed to sit on top of the cage without supervision.* This idea is totally wrong. The parrot may be permanently under stress because of its fear of falling. Sadly, many vets report seeing parrots with split tissue on the breast or abdomen or below the cloaca, as a result of constantly falling on to a hard surface. The pain and suffering which this causes is indescribable. Scabs which do not heal will require expensive veterinary treatment.

- *'My recently weaned young Grey Parrot was so clumsy in flight, I feared it would hurt itself and I had its wings clipped.'* Many of the heavier parrot species (as opposed to parrakeets and lorikeets, for example) are clumsy when they start to fly. They crash-land and bump into objects. But they are learner flyers. This phase lasts only a few days. Clipping their wings is to put them in much graver danger as they are then permanently only capable of crash-landing.

However, I accept that it will not be possible for most people to train an untamed parrot unless it is wing-clipped, but I would suggest that it is allowed to become fully flighted after it has been trained. A gradual clip is recommended: a couple of flight feathers daily over a period of a few days. The flight feathers should not be cut right back. If they are cut very short the irritating feathers shafts could result in the parrot plucking itself. There is too much trauma associated with the instant loss of the ability to fly. If it suddenly loses this power, it will crash or flop on to the floor and could injure itself. A clip which allows it to land with control is recommended. However, care must be taken at all times to prevent escape through a door or a window, as this type of clip does give the parrot some mobility. It must not give the owner a false sense of security.

It is recommended that wing-clipping is not carried out by the owner and, at least initially, not with the owner present. This frightening act could for ever be associated with the owner and may destroy the bird's confidence and trust in that person.

Wing-clipping might make taming a suspicious or wild bird easier. Equally, it might leave a deep scar of mistrust. There are two ways of taming birds and animals. One requires patience and building a rapport with an animal over a long period. It may be many months before any results are seen, but when that animal, in this case a parrot, comes to you voluntarily (something which might happen quite suddenly), it has given you its friendship on its own terms. To me this is more satisfying than the quick fix. It has not been forced to surrender to your attentions. No damage has been done to its ego or its confidence. It does not associate you or the move to your home with being deprived of the power of flight. Alas, today few people have the patience for this approach. The main advantage of wing-clipping in training is that it is easier to train a parrot outside its cage.

Now let us look at some of the assertions about wing-clipping which are usually true.

- *Wing-clipping arrests the physical development of a young parrot.* The pectoral muscles, as well as those of the wings and heart, fail to develop properly. If later the parrot is full-winged, it might be a long time before it can fly well as it will have to build up strength in these muscles. During initial attempts at flight, it will breathe in a very laboured manner.

- A clip which totally deprives a parrot of the power of flight, so that it can only flop on to the ground—with painful results—often results in phobic behaviour. It becomes fearful of everything because it cannot escape. It is totally demoralised and is likely to start to pluck itself.

- A leading parrot behaviourist in the USA states that the average owner of a pet parrot does not have sufficient control over their parrot to be able to handle it if it is not clipped. This would certainly be the case with an aggressive parrot or one which was aggressive for a temporary period while hormones were raging. In this case it would be better to confine the bird to a large indoor flight, where it can fly, or to an aviary, climate permitting. If under normal circumstances they do not have sufficient control over their parrot without clipping its wings, they should not be parrot owners. Or, after the initial wing-clip and training period, they should allow the parrot's flight feathers to grow and be permanently retained. Alan Jones, a leading avian vet in the UK, wrote in *Parrots Magazine*, that wing-clipping 'should not be used as a means of controlling behavioural problems. Most "behavioural therapy" with regard to antisocial animal activities is actually training of the owner in how to cope with a situation correctly rather than treating any disorder on the part of the animal.'

Despite my dislike of wing-clipping, I concede that whether or not they are suitable people to own parrots, many owners would be unable to train a parrot if it was not clipped. If they could not handle it, thereby teaching it discipline, it would fail to become a good pet and could end up as an abused or unwanted bird. The answer is not more wing-clipping but more careful thought—*before* a purchase is made—on what is actually involved in parrot ownership. I also concede that people who live in warm climates, where doors and windows are permanently open, cannot realistically consider keeping a full-winged parrot, so perhaps they might question the wisdom of keeping a parrot at all.

Greg Glendell, of BirdsFirst in Aviculture, states that parrots, with their innate intelligence, do not need to be subjected to the cruel practice of wing-clipping. His words are worth quoting here:

> 'With millions of years of evolution behind them as flying creatures, the decision to clip a parrot's wings should never be taken lightly. In my view it is highly presumptuous of us to assume we can simply take a pair of scissors to such a creature's wings . . . Restricting a parrot's ability to fly should be seen as a measure of last resort, to be used only with great care, when all other 'controls' have failed. Most parrots, given proper training in obedience by someone who understands their behaviour, do not need to have their freedom restricted by such crude methods. The presumption should be that flying birds should be able to fly even when in captivity. Should a fully winged pet bird escape, and it has been properly trained and socialised, it should not be too difficult for the person to whom it is bonded to get it back. Once a bonded bird is 'free', the first thing it will want to do is to get back to its favourite person (Glendell, 1998a).

This statement was borne out by the escapade of a Timneh Grey Parrot in the USA. Hand-reared and affectionate, she was kept in the dining room next to the front hall. There she observed the morning routine which included taking the dogs for a walk. When she was only four months old she was frightened by a new coat that her owner was wearing. Despite the fact that she was wing-clipped, she shot out of the cage into a tree. Then she flew off. The neighbourhood was searched and leaflets were posted regarding her escape—to no avail. There was no sign of her.

Five days later, at eight a.m., Tammy's owner opened the front door to take the dogs out. Suddenly Tammy swooped down, flying straight on to her owner's chest. She had lost one fifth of her body weight. She fed—then slept and slept. Her owner wrote:

> Since Tammy's cage was towards the front of the house, she knew we walked the dogs every morning. I believe from the time Tammy realised she was lost, she flew around the neigh-

bourhood trying to find our house. When she heard the birds screeching in the bird room she was able to locate the house. However, she had to find a way to get back in. Knowing that we walked the dogs every morning, she positioned herself in the crab-apple tree in front of the house (Christian, 1998).

This story perfectly demonstrates the futility of all but the most severe clip—which should never be used on a young bird. It also demonstrates the extraordinary ability which even young parrots have to survive for several days without food, should they escape. I know of several other similar incidents. The main problem in recapturing many escapees is that they have a great fear of flying down.

Question
I obtained my Grey Parrot at the age of 14 weeks from a breeder. His wings were clipped. He is now 14 months old and has had an infection in one wing. It obviously hurt him and he chewed the feathers and the skin. My vet carefully removed the feather stumps but when the new feathers grew, they dropped out when only half formed. Why has this happened and what—if anything—can be done to rectify the problem? My vet said it was a result of wing-clipping but why do other wing-clipped parrots not have this problem?

Answer
A bad clip, too close to the shaft, which crushed some of the feather follicles, was probably to blame. Lack of blood supply to the skin inside the cavity supporting the feather shaft causes the new feathers to fall out before they are fully developed. In a similar case, hormone treatment was successful in nourishing the follicles and restoring the blood supply to the growing feathers. Half a 5 milligram tablet of Ovarid was crushed in a teaspoon and given in liquid twice a week. However, whether this would solve the problem in your parrot would depend on the severity of the condition.

Wing-clipping can also result in an infection around the tail feathers. This happens when the clip is so drastic the bird falls repeatedly so that the tail feathers are broken. The broken shaft is forced into the skin after another fall. The irritation causes the parrot to pluck out the new feathers as they grow. Urgent veterinary attention is needed in cases where wing or tail feather shafts have caused an infection.

Question

My African Grey is 20 months old. Before I bought him I read every parrot book I could find and each one said that it was in the bird's safety to have its wings clipped. I bought my Grey from a local breeder. The bird was parent-reared. After approximately eight weeks he was a tame, loving creature who gave us kisses and would snuggle on my shoulder for a scratch. I was worried that one of my children would leave a door open—and my best mate would be gone. So approximately two months ago I had his wings clipped. I regret it so much. He has reacted to this so badly. How long will it be before his feathers grow back?

Answer

I am very sorry that you followed what has become almost standard advice to have your Grey's wings clipped. How long it will be before his flight feathers are replaced depends on when he last moulted. Not all flight feathers are replaced within one year but enough wing feathers will be replaced for him to fly within 12 months of his last moult. This might occur much sooner.

I commend you for buying a parent-reared Grey and having the patience to tame him. It must be very frustrating for breeding pairs to have all their chicks removed from them at an early age. They need to rear some chicks to independence, in my opinion. As you have shown, it does not take long to tame a parent-reared young one obtained soon after fledging.

Y

Yawning

There is still no widely accepted answer as to why humans yawn. A theory that has found favour in recent years is that yawning cools an overheated brain. One study found that applying cold packs to human heads nearly eliminated yawning, and another that it occurs more often in higher temperatures. It seems likely that the same applies to parrots. However, yawning could be an indication of disease of the upper respiratory tract. Consult a vet and/or supplement the diet with Vitamin A.

EPILOGUE

Thirty per cent of the world's parrot species are in a threat category such as Endangered. The biggest cause of population declines is destruction of habitat and of the larger trees which are used for nesting. The majority of the world's parrots are found in the tropical forests which are being destroyed at the rate of something like 180,000 square kilometres (70,000 square miles) annually. Hunting for trade has endangered the existence of many others.

If you keep a parrot or parrots, spare a moment to think about the thousands of ancestors of your bird that died in trade or were removed from their natural environment for a life in captivity. Without these birds there would be no parrot companions. We owe them a huge debt—and there is a way in which we can repay a little of it. Many parrot conservation projects exist now. Some study the habitat, food resources and nest site availability of certain species to determine the best method of preserving the parrots and their habitat. A number of projects exist to educate people regarding the intrinsic value of the parrots they formerly shot or trapped. Such education projects have proved very effective in reviving the fortunes of a number of critically endangered parrots reduced to a few hundred individuals, such as the Philippine Cockatoo. Many more projects could be equally successful if the money was available to fund them. I would urge all parrot owners and parrot club officials to support such organisations as Fundación Loro Parque (see www.loroparque-fundacion.org), the World Parrot Trust (www.parrots.org) and national organisations that have successful conservations programmes for threatened parrots. These include:

Aquasis (Brazil) www.aquasis.org
Fundacion ProAves (Colombia) www.proaves.org
Fundacion Jocotoco (Ecuador) www.fjocotoco.org

REFERENCES & SOURCES

References

BLACK, M. and S. (2000). A New Life for Louie, *The Grey Play Round Table*, Winter, 10–12.

CHRISTIAN, L. (1998). Tammy come home! *Parrots*, April/May, 42.

CLARK, P. (2000). Humans and Greys . . . the subtler side of things, *The Grey Play Round Table*, Winter, 24–25.

EZRA, A. (1929). Death of a famous Ring-necked Parrakeet, *Avicultural Magazine*, fifth series VII, 3, 58–59.

FOSTER, S. and HALLANDER, J. (1999). The theory of behavioural compatibility as it applies to African Greys and cockatoos, *Pet Bird Report*, 8, 5, 18–23.

GARNETZKE-STOLLMANN, K. and FRANCK, D. (1998) Sibling Relationships as a Means to Acquire Reproductive Ability in Spectacles, *International Parrotlet Society*, VII Nov/Dec., 19–24.

GLENDELL, G. (1998a). Wing-clipping, The debate continues . . ., *Parrots*, April/May, 38–39.

——(1998b) Freddie's story, *Parrots*, June/July, 42–43.

HALLANDER, J. (1999). Congos and Timnehs, is there a difference? *The Grey Play Round Table*, Spring 4, 17–19.

HESS, G. (1951). *The Bird: its life and structure*, Herbert Jenkins, London.

MARSDEN, S.J. (1995). The ecology and conservation of the parrots of Sumba, Indonesia, *PsittaScene*, 7, 2, 8–9.

MEEHAN, C.L., J.R. MILLAM and J.A. MENSCH, 2003, Foraging Opportunity and increased physical complexity both prevent and reduce psychogenic feather picking by young Amazon parrots, *Applied Animal Behavioural Science*, 80: 71–85.

MEEHAN, C.L., J.R. MILLAM and J.A. MENSCH, 2004, Environmental enrichment and development of cage stereotypy in Orange-winged Amazon parrots (*Amazona amazonica*), *Developmental Psychobiology*, 44: 209–218.

ROPER, T.J., 2003, Olfactory discrimination in Yellow-backed Chattering Lories *Lorius garrulus flavopalliatus*: first demonstration of olfaction in Psittaciformes, *Ibis*, 145, 689–691.

SACKS, P., 2014, Various causes lead to dreadful bone disease, *Talking Birds*, September: 14.

SAWKINS, L. (1996–7). Trust takes time, *Just Parrots*, Dec./Jan., 28–29.

SHAY, K.E. (1999/2000). In my experience, *Parrots*, Dec./Jan., 440–42.

SIDEBOTHAM, H.M. (1928). *Round London's Zoo*, Herbert Jenkins Ltd.

SONNENSCHMIDT, R. (1995–6a). Holistic treatment for birds, *Just Parrots*, Dec./Jan., 36–38.

——(1996b). Holistic treatment for birds, *Just Parrots*, April/May, 58–61.

STOLLENMAIER, S., 2006, Pet Grey Parrot Behaviour, in *A Guide to Grey Parrots as Pet and Aviary Birds*, Rosemary Low, ABK Publications.

WAGNER, M. and SONNENSCHMIDT, R. (1997). *Just Parrots*, Feb./Mar., 42–43,

WARREN, D., PATTERSON D. and PEPPERBERG, I. (1996). Mechanisms of English vowel production in a Grey Parrot *(Psittacus erithacus)*, *The Auk*, 113, 1, 41–58.

WHITE, K. (1999). Dinosaur and the sock monster, *Pet Bird Report* 8, 6, 64–65.

ZADALIS, E. (1996). Feather Pickers' Forum, *The Grey Play Round Table*, Fall, 7.

Sources

Auk, The, American Ornithological Union Montana Co-op Wildlife Research Unit, University of Montana, Missoula, Montana 59812, USA.

Grey Play Round Table, The, P.O.Box 190, Old Chatham, NY 12136–0190, USA.

International Parrotlet Society, The, P.O.Box 2428, Santa Cruz, Ca 95063–2428, USA.

Parrots (formerly *Just Parrots*), Imax Visual Ltd, The Old Cart House, Applesham Farm, Coombes, West Sussex BN15 ORP, UK.

Pet Bird Report, The, 2236 Mariner Square Dr 35, Alameda, Ca 94501–1071, USA.

PsittaScene, The World Parrot Trust, Glanmor House, Hayle, Cornwall TR27 4HB.

INDEX

Numbers in bold are main headings

abuse 33–34, 70
adolescence **75–76**, 91
age 21, **143–144**
aggression 22, **76–80**, 84, 141, 145, 157, 162, 188
allergy 38–40, 113,
Amazon, Blue-fronted 41, 89, 95, 98, 134, 155–156, 166, 185, 187
  Cuban 77
  Double Yellow-headed 138
  Orange-winged 27, 33–34, 69, 111
  Vinaceous 69
  Yellow-fronted (Yellow-crowned) 10–11, 161, 163 169
  Yellow-naped 138–139, 165
Amazons 29, 45, 54, 63, 65, 67, 68, 78–79, 81, 91, 111, 131, 138, 146, 147, 148, 150, 156, 157, 161, 162, 184, 186, 188
anthropomorphism **80–81**
arthritis 21, 143
aspergillosis 33
attack 68, 77, **81–85**, 187–188
attention-seeking **42–43**

Bach, Edward 86
back, lying on **87**
bands – *see* rings
bathing **87–89**, 115
beak clicking **89**
  grinding 35, **89**
behaviour, breeding **55-58**, 162–165

hormonal 41–42, 91
  nesting 185
  patterns of **67–69**, 80
  problem **43–45**
  phobic **145–146**
  'reading' **34–35**, 68–69, 122
  sexual 162–165
  stereotypic **173–174**
  undesirable 43–45
biting 16, 81, **89–92**, 166, 198
blindness **92**
blushing **92–93**
boarding **93**
bonding **94–96**, 172
bond, pair **60–61**, 84, 146, 160, 172
branches 102–103, 156, 190
Budgerigar 31, 45, 51, 62, 92, 123, 131, 132, 133, 166, 176–177
buttons **96**

cage, covering the 156
'cage-hate' **97–98**
cages, stacking 106
caiques 63, 67, 161
calcium, deficiency of 19, 132–133, 145
children 16, **99–100**
Cockatiel 19, 21, 45, 51, 52, 58, 62, 63, 64, 66, 113, 123, 129, 131, 132, 133, 143, 157, 164, 181, 183, 184
Cockatoo, Bare-eyed 59, 207
  Citron-crested 87, 122

Cockatoo Bare-eyed – *continued*
  Eastern Long-billed – *see*
    Slender-billed
  Galah 50, 58, 137
  Glossy 51
  Goffin's 103, 124, 192–193
  Lesser Sulphur-crested 61,
    82–85
  Little Corella – *see* Bare-eyed
  Moluccan 27, 43–44, 59, 68,
    126–127, 207
  Palm 59
  Philippine 217
  Slender-billed 136
  Sulphur-crested 57
  Triton 152
  Umbrella 59, 60, 68, 81–82, 100,
    184–185, 188–189, 199, 207
  Yellow-tailed Black 51,64
Cockatoos, black 68
  white 67–68, 80, 100–101,
    102–104, 119, 139, 146 153,
    161, 162, 193
collar 116, 117, 139
colour, influence of 22–23
Conure, Austral 50
  Golden 205
  Green-cheeked 91, 164
  Nanday 116–117
  Patagonian 37, 64
  Sun 127–128
  White-eyed 57
conures 45, 49, 63, 66, 67, 81, 117,
  128, 146, 161, 162
coughing **100**, 158, 170
'cuddly-tame' **100–101**

deficiencies, dietary 113, 116, 138
destructiveness **102–105**
diet 19, 115
discipline – *see* training
disease, psittacine beak and feather
  113
diseases 33, 42, 158, 171, 191
dominance 22, **105–106**
dunking 119

egg-binding 132

enrichment, environmental 26–27
environment **21–22**, 58
  emotional **24–26**
euthanasia **107**
eye contact **107–108**, 141

fear **109–111**, 145, 157
feeding **51–53**, 54–55, **117–119**,
  138
feet, biting at – *see* toes, mutilation
  of
flight **64–65**

gender **31**
gland, preen (uropygial) 147–148
glaring 107–108
gloves 141
grieving **120**
growling 34, 111

handling 43, 68, 79, 94, **121**
hand-rearing 15, **19–21**, 40, 53
hands, fear of **122–123**
health 19, **42**
hormonal influences **41–42**
'house-training' **123**
hypocalcaemia – *see* calcium,
  deficiency of

importation 31–33
independence **58–60**
infection, sinus 171
influenza, avian 32
intelligence **124–125**

jealousy 80, 112, **126–127**, 156, 157,
  162

Kakariki, Red-fronted 50, 62, 66
Kea 50, 125, 136
kissing **129**

lameness **130–131**
laughing **131**
laying **131–133**, 162–163
lice 112
loneliness **133–134**
longevity 143

Lorikeet, Green-naped 160
  Little 64, 110
  Musk 110
  Musschenbroek's 65
  Stella's 18, 29, 52, 66
Lory, Black 31
  Black-capped 159, 162
  Chalcopsitta 31
  Dusky 26
  Rajah 18, 23, 69, 110
  Yellow-backed 169
lovebirds 31, 45, 62, 63, 66

Macaw, Blue and Yellow 9, 32, 69,
    92, 105, 124, 143–144, 203
  Green-winged 57, 60, 105, 151
  Hyacinthine 50, 54, 57
  Lear's 55
  Scarlet 70, 144
  Severe 57
macaws 45, 57, 58, 63, 66, 67, 87,
    96, 99, 102, 119, 123, 137, 146,
    152, 153, 170, 199
medicine, traditional Chinese
    184–185
microchips 152, 153
mimicry **135–136**
mites 67, 112
moodiness 16, 17–18, **137**
moulting **137–138**
movements, flight intention
    110–111
music **138–139**
mutilation, self **139–140**, 191–192

nervousness 141
nest sites 56–58
net, catching 141

obesity **142**
operant conditioning 28–29
origin **31–34**
oviduct, prolapse of 163

parasite, *Giardia* 113
Parrakeet, Antipodes Green 50
  Austral *see* Conure, Austral
  Barraband's 64

  Blue-winged 50, 57
  Brotogeris 57, 148
  Hooded 52
  Neophema 52, 66
  Paradise 52–53
  Princess of Wales 64
  Redrump 52
  Ringneck 146, 202
  Rosella 57
  Splendid 52, 64, 66
  Swift 64
  Turquoisine 52
Parrot, Amazon – *see* Amazon
  Blue-winged – *see* Parrakeet
  Caique *see* Caiques
  Cape 51
  Eclectus 31, 38–40, 55–57, 58,
      67, 112, 115, 120, 146
  Great-billed 67
  Grey 16, 17–18, 19, 25–26, 27,
      31, 32, 41–42, 45, 49, 50, 52,
      63, 65, 66, 67, 68, 75–76,
      78–79, 87, 88, 91, 94–95,
      98–99, 106, 107, 108, 111,
      112, 113, 114–115, 121, 123,
      125, 126, 131, 135, 139, 155,
      161, 164, 166, 168, 171, 173,
      176, 181, 182–183, 197, 200,
      201, 206, 209, 212–213, 214
  Grey-headed 152, 153
  Ground 52
  Hanging 50, 53
  Hawk-headed 69, 139–140
  Jardine's 66
  Mallee Ringneck 52
  Massena's 181
  Meyer's 191
  Pesquet's 51
  Pionus 31, 36–37, 58, 63, 148
  Pionus, Blue-headed 33–34, 66
  Red-bellied 66, 95, 199
  Red-cheeked 56
  Senegal 66, 75, 137, 202
  Superb *see* Barraband's
      Parrakeet
  Vasa 49, 174
Parrotlet, Spectacled 172
parrotlets 63, 90, 123, 177, 188

parrots, wild-caught 31–34, 77
PBFD *see* disease, psittacine beak
    and feather
peanuts 39–40, 142, 182
pellets 19, 39, 132
perches 105, 143, 190
plucking, feather 42, 45, 100, 111,
    112–117, 174
poisoning, heavy metal 113,
    194–195
polyoma 113
predictability **148**
preening 34, **146–148**
prevention **149**
punishment **149**

reinforcement, positive **27–28**
regurgitation **150–151**
rehoming 44
remedies, Bach **86–87**, 115, 141
    homoeopathic 86–87, 115 141
rings 111, **151–152**
rickets **130–131**

salmonellosis 33
scratching (on floor) 98–99
screaming 17, 41, 42, 134, **154–162**
sexing, DNA 117, 128
sexuality **165–166**
shoulders, sitting on **166–167**
shower 88, 137
singing 138–139, 161
sleep **168–169**
smell, sense of **169–170**
smoking **170–171**
sneezing **171**
stepping up 28, 90 114, 122, 183,
    196–200
socialisation **30–31, 171–173**
speech, the mechanism of **177–178**

spoon-feeding 20
stand 102, 104
stress 111, 112, 116, 139, **157**, 206
stroking 194
sun **174**
swings 37, 105–106

tail-bobbing **176**
talking **176–181**
taming **181–184**
teasing 99
television 116, 131, 146, 158, 168,
    185–187
territoriality 80, 82, **187–189**
toenails **189–191**
toes, mutilation of **191–192**
tool-use **192–193**
touching **193–194**
toys 26–27, 104, 134, 154,
training 69, 82, 90, 154, 194–195,
    195–201
tricks **201–203**

veterinarians 42, 43, 86–87,
    112–113, 157, **204**
vision, avian 187
    colour 204–205
vitamin A, deficiency of 113, 116,
    171, 206
vitamin D^3 19, 130, 131, 132
vocalisations **65–66**, 158, 159–161

weaning 58–59, 60, 76–77, 78,
    206–207
wheezing **206**
whining **206–207**
wing-clipping 79–91, 110, 145,
    207–214

yawning **215**